ARIE ESKINAZI

Hook. Edit. Convert.

A Paid Social Video Editor's Playbook to Mastering Scroll-Stopping Ads and Building a Creative Career

vincivision

First edition

ISBN (paperback): 979-8-9993202-0-9
ISBN (hardcover): 979-8-9993202-1-6

This book was professionally typeset on Reedsy.
Find out more at reedsy.com

Contents

Preface

How to Use This Book: This book is built around the 3 key phases of becoming a high-performing Paid Social Video Editor:

Part I: Foundations (Chapters 1–4)

Build your mindset and systems. Learn scroll psychology, ad archetypes, and how pros stay fast and organized.

Part II: Execution & Creative Craft (Chapters 5–12)

Structure, strategy, and editing techniques. Understand metrics, platform differences, and how to build converting ads — including advanced UGC and the role of tone and clarity.

Part III: Experimentation, Business & Scaling (Chapters 13–18)

Test like a scientist. Build your portfolio. Choose your path: grow in-house, freelance, or launch your own agency.

You don't have to read it in order. Jump to what you need. This is your reference manual — screenshot it, revisit it, apply it. Let's get to work.

All references and sources cited throughout this book are listed in full at the end in the "Sources" section, organized without inline markers for cleaner readability.

Introduction

Why Paid Social Video Editing Is the New Creative Goldmine

The demand for video has exploded — but not just any kind of video. We're in an era where short-form, performance-driven content is king. TikTok, Instagram Reels, YouTube Shorts, and UGC ads have changed the way people consume information and how brands capture attention.

And this shift has created a golden opportunity for editors.

Traditional editing focused on polish. On storytelling. On perfection. But paid social? It's built on speed, experimentation, attention, and conversion. It's scrappy. It's strategic. It's full of data. And it needs editors who get it.

This book is your blueprint — whether you're brand new to editing or already experienced and looking to break into the paid social space. You'll learn how to:

- Craft ads that grab attention and drive clicks
- Set up professional workflows like top creative teams
- Master creative testing and data-driven iteration
- Work with agencies and brands — or go solo
- Scale your career from freelancer to creative strategist

If you know how to cut a basic video, you're ready for this book. This isn't a course on Premiere — it's a course on performance.

I started out just like many of you — as a video editor with creative instincts. Over time, I became obsessed with performance: Why did one edit work while another flopped? How could I make videos that didn't just look good — but sold products? That obsession led me to where I am today: producing ads for 7- and 8-figure brands, leading creative strategy, and running my own video business.

This book isn't about fluff or theory. It's a practical playbook built on real client work, real ads, and real metrics. If you want to get in, stand out, and grow fast — you're in the right place.

I

Part I: Foundations

Build your mindset and systems. Learn scroll psychology, ad archetypes, and how pros stay fast and organized.

1

Beginner to Pro – Do You Need to Be a Master Video Editor?

Here's the truth nobody tells you: you don't need to be a "pro editor" to start working in paid social.

In fact, some of the highest-performing ads on Meta and TikTok look like they were made in 15 minutes on a phone — because sometimes they were. The difference is in the strategy, not the flash.

What Actually Matters

- **The Message:** Can you communicate the value of a product clearly in under 15 seconds?
- **The Hook:** Can you stop someone mid-scroll?
- **The Retention:** Can you hold their attention just long enough to deliver a CTA?

These are the real skills that make you valuable. Not transitions. Not motion graphics. Not fancy LUTs.

If you have a basic grasp of editing software — like Premiere Pro, CapCut, Final Cut, or even mobile apps — you can already start building ads that outperform big-budget campaigns. It's not about expensive gear or cinematic shots. It's about clear messaging, a good hook, and making something feel native to the feed.

You'll absolutely get better over time — and you should care about clean edits and pacing. But don't let perfection stop you from starting. **Paid social rewards editors who can move fast, iterate often, and think like marketers.**

Learn as You Go (With Real Data)

One of the best things about this niche is how quickly you learn what works. Every campaign gives you feedback:

- Did the audience drop off at 3 seconds? Try a different hook.
- Did the cost per click drop on Version B? Double down on that format.
- Did your "ugly" edit outperform the polished one? Good. Now you know.

In traditional film editing, it might take months to get feedback. In paid social, it's instant — and it makes you better with every ad you cut.

You're Not Behind — You're Early

Right now, most video editors are still focused on wedding videos, YouTube vlogs, or narrative short films. That's fine. But the editors who learn how to create content that sells — and how to think like marketers — will always be in demand.

You can start right now:

- Use a client's existing footage to build a UGC-style ad.
- Recut content into vertical formats.
- Apply creative formulas you'll learn in this book — and get results faster than you think.

This chapter isn't here to hype you up — it's here to show you the truth: paid social editing is not about being a polished artist. It's about being a performance-minded creative, and the bar for entry is lower than you think.

2

Understanding the Scroll & Creative Archetypes

Before you make a cut, before you drop footage on a timeline, before you even open Premiere — you need to understand one thing:

The scroll is the battlefield.

You're not competing against other ads. You're competing against everything — memes, pets, thirst traps, news, creators, chaos. If you don't grab attention instantly, you're invisible.

And the scroll is ruthless. You get maybe 1.5 seconds. Maybe.

But here's the good news: **there's a science behind what makes people stop.** Once you understand it, you'll stop editing blindly — and start crafting content that gets noticed.

What Makes a Thumb-Stopping Hook?

People stop for content that:

- Surprises them
- Shows them something relatable
- Sparks curiosity
- Solves a problem they care about
- Feels native to the platform

This isn't guesswork. It's human psychology. Our brains are wired to respond to novelty, emotion, pattern breaks, and relevance.

A strong hook can be:

- A bold question: *"Why is nobody talking about this?"*
- A pattern interrupt: *A guy in a bathrobe in a Walmart parking lot*
- A jump cut to a pain point: *"My skin was freaking out until I tried this"*
- A visual oddity: *zoomed-in ASMR cleaning footage with text overlay*

You're not here to be subtle — you're here to **interrupt passivity**.

Emotions That Stop the Scroll

Certain emotional triggers consistently outperform:

- **Curiosity** → *"I didn't believe this until I tried it"*
- **FOMO** → *"Everyone is switching to this one app"*
- **Frustration/Relatability** → *"This used to take me 3 hours... now it's 15 minutes"*
- **Shock or Intrigue** → *"This ingredient is banned in 3 countries"*

9

· **Humor** → *"My grandma tried this and now she thinks she's an influencer"*

The emotion is the vehicle. The product is the passenger.

Always lead with what makes people *feel* something.

The 5 Core Paid Social Ad Archetypes

These aren't just ad formats — they're frameworks for persuasion. Knowing these archetypes lets you speak the language of performance ads fluently, helping you frame any brand or offer in a way that resonates with your audience.

Before diving in, here's a quick preview of the five core archetypes:

1. Classic Ad Structures
2. Informational Content (Masquerading as Value)
3. UGC Testimonial Ads
4. Visual / Window Shopping Formats
5. Story Variations & Bulletins

1. Classic Ad Structures

These are direct-response staples. They look clean, structured, and message-first. Usually feature white graphic backgrounds, centered or justified copy, and clear CTA moments. They feel like minimal landing pages in motion.

a) Problem–Solution
Starts by naming a pain point the audience already feels.

- Hook: taps into frustration, confusion, or a fear of missing out.
- Structure: the problem is quickly followed by a product shown as the solution.
- Style: no visual clutter — often white background, clean fonts, product demos.

Example Headlines:

- "Hate waxing?"
- "Still using drugstore shampoo?"
- "Tired of meal planning every week?"

This format works because it gets straight to what the viewer cares about and offers an immediate payoff.

b) Desirable Claim / Comparison

Leads with a bold claim or benefit. Then contrasts it with an inferior alternative.

- Hook: a visual or headline that signals a major upgrade or discovery.
- Structure: new vs. old, better vs. worse.

Example Headlines:

- "Why I switched from X to Y"
- "This $12 serum beats $100 brands"
- "$10 Boots?!"

This format drives curiosity by implying the viewer is missing out.

c) Classic UGC Hybrid

Mixes clean, graphic layouts with raw testimonials or voiceovers.

- Structure: intro hook using UGC, followed by cleaner text-driven product section.
- Feel: combines the trustworthiness of UGC with the polish of graphic ads.
- Works well when testimonial clips are strong but need structure or pacing.

Great for brands that want authenticity without sacrificing control over message.

2. Informational Content (Masquerading as Value)

These look like native content pieces — think listicles, infographics, or social explainers. They delay showing the product to build trust and curiosity first.

a) Reasons to Quit / Switch

Starts with a provocative list: "3 Reasons Why..." or "You've Been Doing X Wrong"

- Hook: a numbered list or a controversial claim.
- Style: clean titles, hate-driven opening, followed by persuasive logic.

Examples:

- "Why your toothpaste is ruining your teeth"
- "3 Reasons People Are Quitting X"

This format taps into pain, logic, and curiosity in quick succession.

b) Easy Trick

A one-line "life hack" style hook with a twist.

- Structure: show something simple and relatable, then offer a better way.

Example:

· "Easy trick to cut calories without skipping meals."

It feels like advice — not a pitch. That earns trust.

c) Myths / Misconceptions

Breaks what the viewer believes to be true.

· Hook: "Myth:" or "You've been told..." setup.
· Strategy: debunk, then replace with your product.

Examples:

· "Myth: You can't eat carbs and lose weight"
· "Myth: You need 8 hours of sleep every night"

Creates tension — then relieves it with the product.

d) Quiz-Style / Interactive

Presents choices as if the viewer is playing a quiz.

· Structure: "What's your skin type?" → multiple answers → same product path.
· Great for wellness, beauty, self-care, finance.
· Makes the ad feel interactive, even though it's passive.

Use clear overlays and short questions to lead viewers down a funnel.

e) How-To / Step-Based

Tutorial-style ads with a straightforward, educational tone.

· Hook: "How to..." or "3 Steps to Fix..."
· Structure: clear, numbered steps. Easy to follow.

Example:

· "How to fix teeth that move after braces"

Perfect when your product solves a common but overlooked issue.

3. UGC Testimonial Ad

Unpolished, human, and real. These are trusted because they don't feel like ads — they feel like advice from a friend.

a) Personal Context On-Ramp

Opens with a relatable story or moment of vulnerability.

· Hook: "I used to..." or "The first time I tried..."
· Sets up empathy and interest before shifting into product talk.
· Often paired with soft music, natural light, and real backgrounds.

Examples:

· "I've always struggled with acne..."
· "Trying new makeup is too expensive."

This type of UGC builds trust and emotional connection fast.

b) Experiential Walkthrough

Shows someone using the product while talking about it.

· Feels like a casual review or personal vlog.
· Focus is on demonstrating and describing benefits.
· Usually voiceover + b-roll or selfie-style explanation.
· Less storytelling, more usage-based proof.

Works best when the product itself has strong visual interaction or visible results.

4. Visual / Window Shopping Archetypes

These are built to stop scrollers with high visual impact. Think TikTok-native energy. Short captions, flashy visuals, price callouts.

a) Visual Problem–Solution
Before/After split screen with a reveal.

- Structure: problem shown side-by-side with solution.
- Great for beauty, cleaning, fitness, transformation.
- Text often minimal — visuals do the work.

b) Visual Problem Solvers
The entire video *is* the problem being solved.

- Example: messy fridge → satisfying restock with product.
- Viewer gets the message without a single word.

These are great for sound-off environments or scroll-heavy platforms.

c) Window Shopping / Deal Reel
Fast montage of deals or products.

- Caption overlays: prices, short comments, reactions.
- Style: "$10 finds at Target" / "Amazon must-haves"
- TikTok native and highly relatable.

Use punchy music, transitions, and text to sell the aesthetic.

d) Hype Reel

Looks like a movie trailer or product teaser.

· Cut fast, dramatic music, animated graphics.
· Great for tech, new launches, or brand awareness.

You don't sell benefits here — you sell *vibe*.

5. Story Variations & Bulletins

These are hybrids — creative twists on the core formats.

a) Adapted Archetypes
Mashups of multiple formats.

· Example: Start with personal story → cut to "3 Reasons Why" → end with visual testimonial.
· Often short, stylized, or comedic.
· Great for experienced editors who want to push the envelope.

These reward experimentation — and often outperform when done right.

b) Bulletin Style
Like a split-screen infographic with text info at the bottom.

· Top half: person talking or footage.
· Bottom half: bullet points or step-by-step info.
· Looks like a visual whiteboard.
· Common in wellness, skincare, finance, and explainer videos.

Archetype ≠ Template

These aren't checklists — they're **starting points.**

The best ads *combine* elements: a funny testimonial with a visual demo. Or a bold explainer that ends in a product transformation. Think of these as creative skeletons — your job is to give them life, voice, and rhythm.

How to Use Archetypes in Real Projects

Ask yourself:

- What emotional lever is best for this product?
- Can I show value visually or narratively?
- Is this ad about awareness, education, or conversion?

And remember:

- The right structure depends on the audience and platform.
- What works on TikTok may flop on Meta.
- What works for a $10 impulse buy may not work for a $300 skincare bundle.

3

Prepping for Success — Job Setup, Naming & Project Structure

Part 1: Creating Your Client Folder System

Before you ever drop footage into Premiere or brainstorm a hook, you need to do something that will save you hours of confusion, backtracking, and frustration:

> ☞ **Organize your files. Like a pro.**

Step 1: Create a Master Client Folder

Start by creating a master folder named after your client or the brand. Example:

```
/Clients/GlowSkin
```

Everything you create for this brand — past, present, and future — will live inside this folder.

Step 2: Inside the Client Folder, Create Two Subfolders:

1. **ApprovedAssets** – This is where all brand-provided assets go
2. **Jobs** – This is where all your editing projects will live

What "Jobs" Really Means

In this system, we use the term **Job** to define a single creative concept.

Each Job can — and should — contain **multiple ad variations** (at least 4 is the standard).

A Job is not just a "video." It's a campaign test, a creative hypothesis, a performance container.

Now Let's Build the "ApprovedAssets" Folder

Your full structure should look like this:

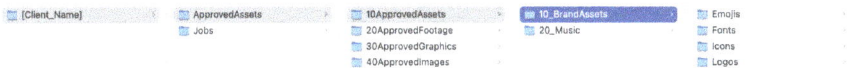

Important: There are two "ApprovedAssets" levels in this structure.

The top-level **"ApprovedAssets"** folder contains:

- 10_ApprovedAssets
- 20_ApprovedFootage

- 30_ApprovedGraphics
- 40_ApprovedImages

Inside the **"10_ApprovedAssets"** folder, you'll create two additional sub-folders:

- 10_BrandAssets
- 20_Music

What Each Folder Is For:

10_ApprovedAssets → 10_BrandAssets

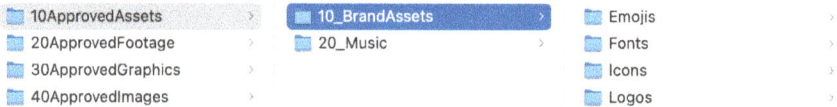

10ApprovedAssets	>	10_BrandAssets	>	Emojis	>
20ApprovedFootage	>	20_Music	>	Fonts	>
30ApprovedGraphics	>			Icons	>
40ApprovedImages	>			Logos	>

This folder contains brand identity essentials:

- Logos (horizontal, vertical, white, black)
- Icons and graphics
- Brand fonts (if custom)
- Emojis (less relevant now as most editing tools have these built-in)

10_ApprovedAssets → 20_Music

A single folder for all client-approved or licensed music. This avoids digging through downloads or losing track of what's cleared to use.

20_ApprovedFootage

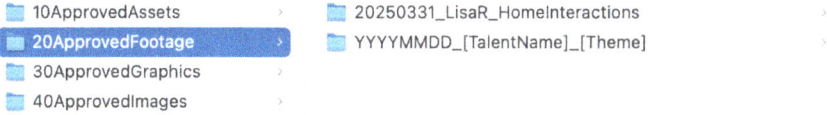

📁 10ApprovedAssets	›	📁 20250331_LisaR_HomeInteractions	›
📁 **20ApprovedFootage**	›	📁 YYYYMMDD_[TalentName]_[Theme]	›
📁 30ApprovedGraphics	›		
📁 40ApprovedImages	›		

Arguably the **most important folder**. This holds:

- All raw footage provided by the brand
- UGC clips, interviews, testimonials, b-roll

You will pull from this folder across multiple jobs — **never copy footage into each job folder.** That bloats your system and causes sync errors when working with teams.

> 💡 **Naming Format:**
>
> Use *YYYYMMDD_TalentName_Theme* - to name each batch of footage:
>   Example: **20250331_LisaR_HomeInteractions**
>
> This gives you instant clarity on the source, date, and context.

30_ApprovedGraphics

📁 10ApprovedAssets   ›   📁 Outros   ›
📁 20ApprovedFootage   ›
📁 **30ApprovedGraphics** ›
📁 40ApprovedImages   ›

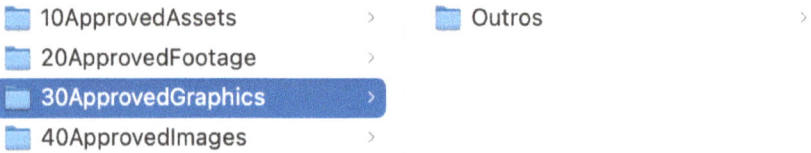

Here's where you store:

- CTA overlays
- "Shop Now" buttons
- Outro animations or branded end cards

Many agencies or brands update graphics frequently. Keep both **old** and **new** versions in here — label clearly.

40_ApprovedImages

Photoshoots, talent photos, product stills, thumbnails, packaging imagery, etc.

Part 2: Setting Up the Jobs Folder & Naming System

Once your **ApprovedAssets** folder is ready and clean, it's time to set up your Jobs system — this is where the actual editing and creative magic happens.

What's a "Job"?

In your workflow, a **Job** is more than just a video. It's a **creative concept**, a campaign idea, and a performance experiment.

Each Job should contain **at least 4 variations** (more on that later), and everything associated with that concept — footage pulls, project files, exports — should be **fully contained** inside the Job folder.

> 💬 Some agencies or brands might call this a "Creative." Unless you're following a client's exact naming convention, use **Job** — it's scalable, intuitive, and cleaner for organizing performance campaigns over time.

Create a Template Job Folder

Start by building a folder named:

```
J#_Template
```

This will act as your master structure to duplicate every time you start a new job.

Inside this folder, create the following structure:

What Each Folder Does:

10_Output

J#_Template	>	10_Output	>	4-5	>
		20_Projects	>	9-16	>
		30_JobAssets	>		

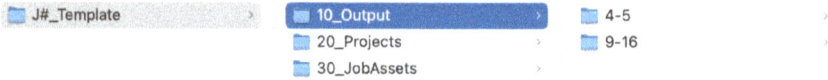

*This is where you save **final exported ads** and variations.*

You can split into folders by aspect ratio (e.g. *[4-5]*, *[9-16]*) or by platform/-client — whatever keeps your exports organized.

20_Projects

J#_Template	>	10_Output	>	Client_Template.prproj
		20_Projects	>	
		30_JobAssets	>	

*This holds your **Premiere Pro project file**, and any others (After Effects, DaVinci Resolve, etc.).*

Start with a clean template project named something like:

```
Client_Template.prproj.
```

> **Tip:** Some editors also include presets, adjustment layers, and branded captions in their template project to save time.

30_JobAssets

📁 J#_Template >	📁 10_Output >	📁 10_Video >
	📁 20_Projects >	📁 20_Audio >
	📁 30_JobAssets >	📁 30_Images >
		📁 40_Other >

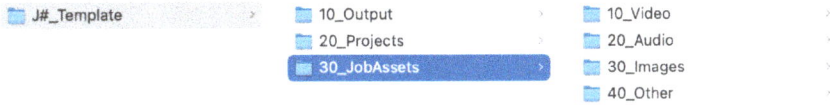

*These are the **specific assets used only for this Job** — not pulled from general ApprovedAssets.*

Inside you'll have:

- **10_Video** → Unique b-roll, stock clips, screen recordings
- **20_Audio** → Custom VO, sound effects, audio repairs
- **30_Images** → Social screenshots, memes, visual aids
- **40_Other** → Text docs, scripts, client notes, metadata

This setup keeps each job self-contained and avoids cluttering your ApprovedAssets folder with one-off elements.

Naming Your Jobs

Now let's talk naming conventions. A well-named job tells you:

- The **job number**
- The **archetype** (UGC, Informational, Visual, etc.)
- The **talent name or main character**
- The **theme or concept**
- Optionally, a short tag for the brand/client or featured product

Example:

25

```
J5_UGC_MarioB_NewToothbrush_VV
```

Breakdown:

- **J5** → Fifth job with this brand
- **UGC** → The archetype used (can also be Classic, Informational, Visual, etc.)
- **MarioB** → Talent name
- **NewToothbrush** → Theme or product concept
- **VV** → Short for VinciVision (client/brand tag)

⚠ Note: If you're working with an agency that already uses Job numbers, you may have to start at **J53** or wherever they left off — always confirm before naming.

Alternate Endings:

If you are already working for a brand, you may not need to include their name at the end of the Job title. Instead, you can use a short abbreviation of the **product being featured**.

For example:

```
J1_UGC_AhamdB_TWK
```

Here, **TWK** stands for *Tooth Whitening Kit.*

A Typical Jobs Folder Might Look Like This:

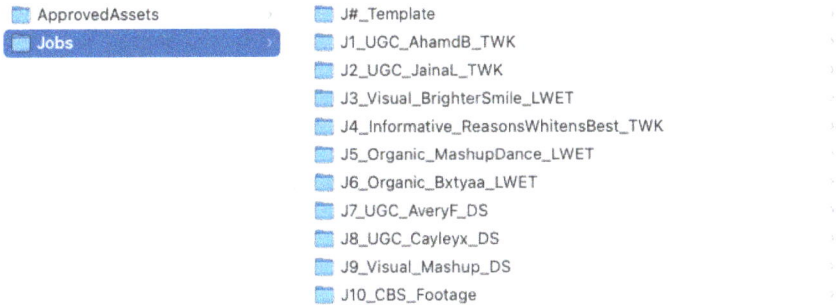

ApprovedAssets	J#_Template
Jobs	J1_UGC_AhamdB_TWK
	J2_UGC_JainaL_TWK
	J3_Visual_BrighterSmile_LWET
	J4_Informative_ReasonsWhitensBest_TWK
	J5_Organic_MashupDance_LWET
	J6_Organic_Bxtyaa_LWET
	J7_UGC_AveryF_DS
	J8_UGC_Cayleyx_DS
	J9_Visual_Mashup_DS
	J10_CBS_Footage

Once the naming system is set, everything becomes easier:

- File sharing with editors is clean
- Export naming becomes more accurate
- You know what's been shipped and what's pending
- You can identify performance concepts at a glance

Part 3: <u>Setting Up Your Project in Premiere Pro</u>

After creating your Premiere Pro project, the next step is critical: **organizing where all your files will live inside your project panel.**

Think of Premiere like your kitchen — your creative lab. If everything isn't organized, you'll waste time hunting for files, rewatching footage, or worse: exporting the wrong version. This structure isn't just for you — it's for the **other editors, supervisors, and clients** who may jump into your project.

Your Two Main Folders: Assets and Sequences

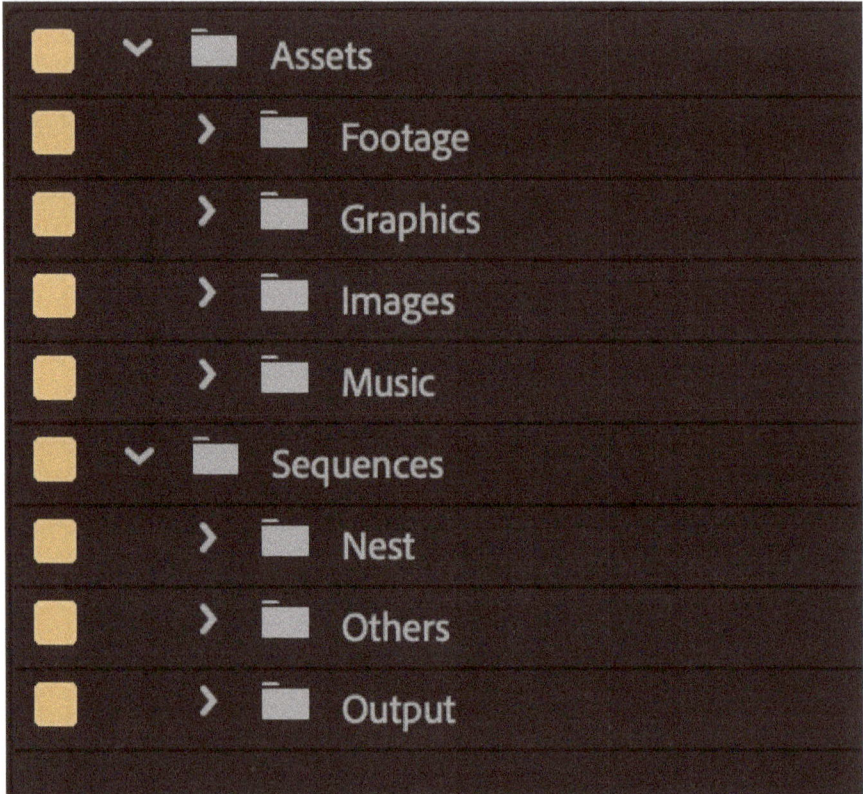

These are standard across the industry. Every editor, no matter the project type, should begin with these

Assets Folder

The **Assets** folder is where all the raw ingredients go — what you pull from to create the final ad. Inside it, you'll create these subfolders:

- **Footage:** All the footage specific to this job.
- **Graphics:** Outros/endcards, CTAs, text style templates, adjustment layers, brand overlays.
- **Images:** Product shots, user screenshots, branding stills, UI captures, etc.
- **Music:** The soundtracks, licensed audio, or recurring musical cues you'll test.

Optional Subfolders You Might Add:

- **Audio:** For voiceovers or soundbites you clean or edit separately.
- **Transitions or Effects:** Drag-and-drop presets or downloaded plugins.

Sequences Folder

This is where the timeline work happens. How you structure this folder can make or break your ability to iterate quickly and collaborate cleanly. Here's how to set it up:

Nest: Contains your nested sequences — modular chunks like intros, problem/solution sections, testimonials, etc. Nesting allows you to make changes once and have them ripple across all ad variations.

- Let's say you exported 4 variations and realize one of the shots you used is off-brand or got flagged — maybe a talent uses the product incorrectly, or there's a distracting background. If that shot is inside a nested sequence, you only need to fix it once, and it updates across all the variations automatically.

- Nesting also helps you visually organize the parts of an ad and build variations with less clutter.

Others: A space to customize how you work.

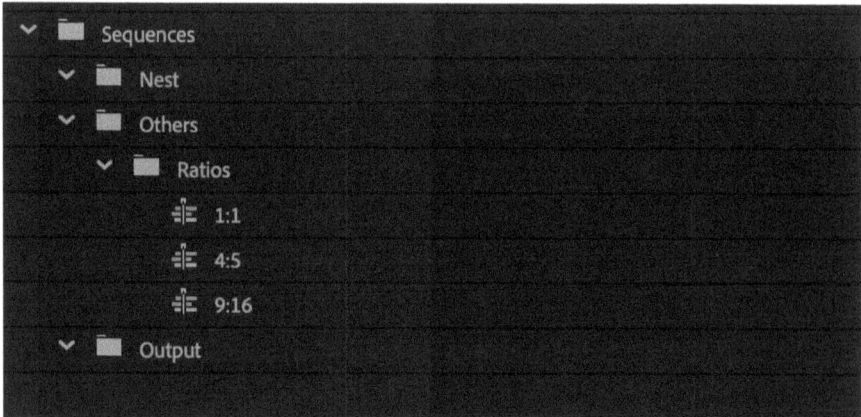

Ratios. Here's where you keep aspect-ratio-specific templates:

```
1:1  (1080x1080)
4:5  (1080x1350)
9:16 (1080x1920)
```

You might also include:

- **Selector:** A sequence where you drop your favorite footage moments — this helps you make creative choices without rewatching 45 mins of footage every time.

Output: Where final cut sequences live, ready to export. Keep this clean — only final versions should be here.

This folder structure keeps your Premiere project **efficient, fast to load, and**

easy to hand off — and it makes collaboration much easier when you're working with agencies or clients giving feedback on multiple versions.

Part 4: Naming Conventions & Structuring Variations

Once your sequences are set up inside your project file, it's time to dive into the **Output** folder. This is where your final variations will live. I'm going to break down not only how to structure these, but also what the naming convention means and why it matters — especially when working with big clients or agencies who need a lot of versions.

Inside the "Output" folder of your Premiere project, you're going to organize your exported sequences. I've included three real examples here to help you visualize this setup.

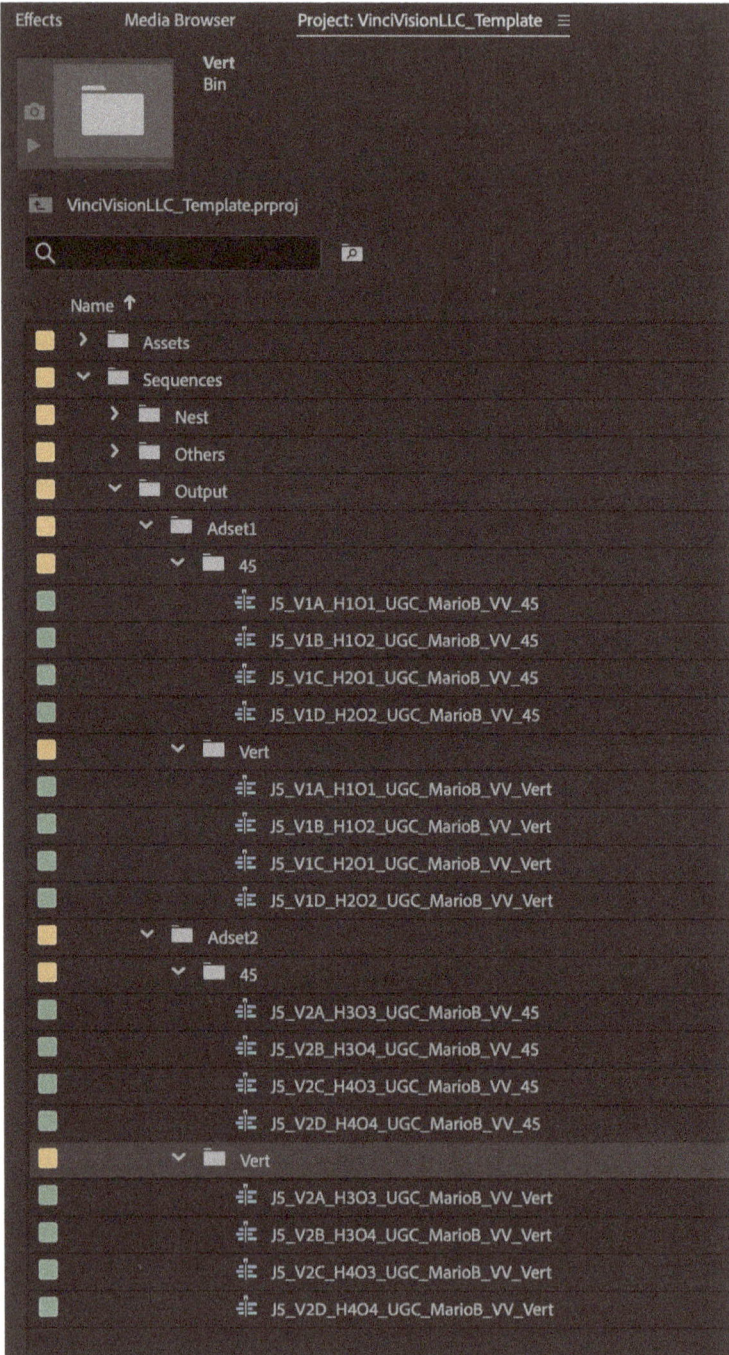

Example 1

You'll notice the **Output** folder includes subfolders organized by **Adsets**. Each **Adset** contains tests with different creative combinations. Inside each Adset folder, you'll find subfolders labeled:

```
45 → For 4:5 aspect ratio versions (1080x1350)
Vert → For 9:16 vertical versions (1080x1920)
You might also see SQ for 1:1 square formats (1080x1080)
```

> **Pro Tip:** Always get the final variation fully approved before building versions in other aspect ratios—otherwise, you'll be duplicating effort across multiple sequences.

Inside those are the actual sequences representing each variation of the job. In large-scale jobs—especially for big brands or agencies—it's common to create more than 4 variations, sometimes up to 16. Why? Because they have the budget to run more creative tests and identify what works best. But generally, 4 variations in one adset is the standard minimum.

Each **"Adset"** represents **one test group**. Within it, variations test different combinations of creative elements.

Naming Convention Format

This naming system isn't something you'll find universally across all brands or agencies, but in my experience, it's the clearest and most scalable structure out there:

```
J#_V#A_(VariationChanges)_Archetype_TalentName or Theme_Client or
Product_AspectRatio
```

Example:

```
J5_V1A_H1O1_UGC_MarioB_VV_45
```

- **J5** – Job number 5 for this client
- **V1A** – Adset 1, Variation A
- **H1O1** – Header 1, Opener 1 (the specific creative test components)
- **UGC** – Archetype (User Generated Content)
- **MarioB** – Talent or actor featured
- **VV** – Short for VinciVision (optional, if you're already in the client project)
- **45** – Aspect ratio (4:5)

In *Example 1*, you'll notice that **only the variation letters (A, B, C, D)** and **header/opener numbers** change within each Adset when testing.

- J5_**V1A**_**H1O1**_UGC_MarioB_VV_45
- J5_**V1B**_**H1O2**_UGC_MarioB_VV_45
- J5_**V1C**_**H2O1**_UGC_MarioB_VV_45
- J5_**V1D**_**H2O2**_UGC_MarioB_VV_45

We're testing how **Header 1** performs with different Openers and how those compare to **Header 2** with the same **Openers**. 4 variations allow us to gather real insights. Maybe **H1** has strong thumbstop performance, but pairing it with **O2** drops conversion. This method helps us narrow down what's actually driving performance.

In **Adset 2**, we might be testing completely new hooks:

- J5_**V2A**_**H3O3**_UGC_MarioB_VV_45

- J5_**V2B_H3O4**_UGC_MarioB_VV_45
- J5_**V2C_H4O3**_UGC_MarioB_VV_45
- J5_**V2D_H4O4**_UGC_MarioB_VV_45

We can then compare **Adset 1 vs. Adset 2** to determine which **Header/Opener family** drives better performance. That means the results from **J5_V1C** and **J5_V2A** (if they're the best performers from each Adset) become your top contenders for future combinations.

This process gives editors and strategists powerful insight into creative impact.

Understanding the Other Variables

Beyond **H (Header)** and **O (Opener),** other common elements you'll test include:

- **R** = **Onramp** / personal context / problem identification
- **S** = **Sales Sequence** (the product explanation and pitch)

Let's look at some examples:

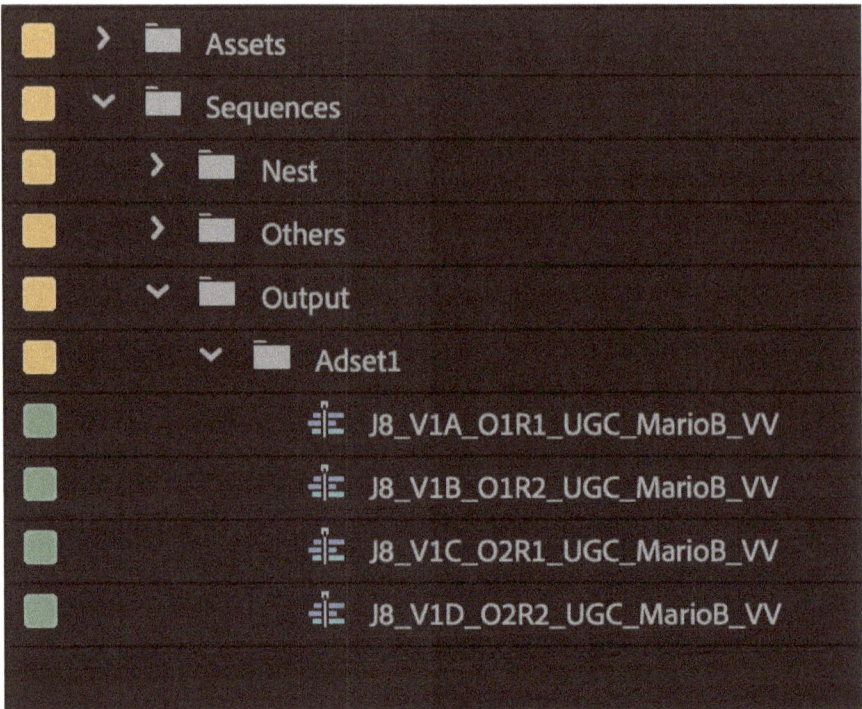

Example 2

In *Example 2*, instead of headers, we're testing **Openers (O)** and **Onramps / Personal Context / Problems (R)**. It's another creative angle to see which openings connect better.

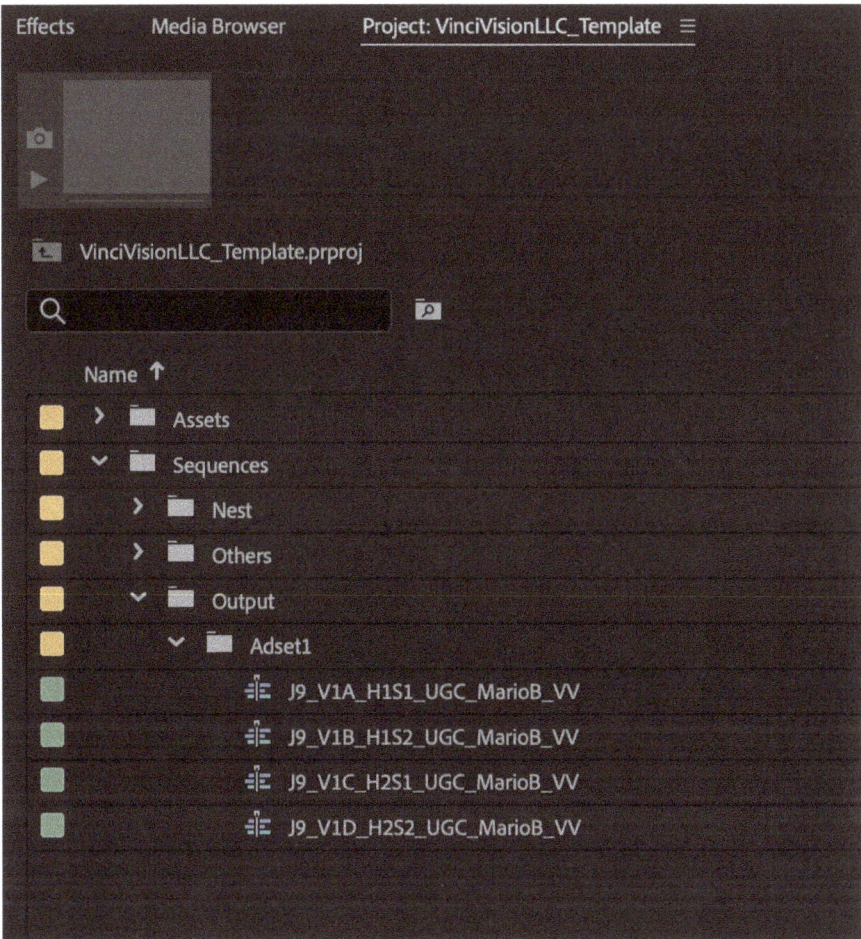

Effects   Media Browser   Project: VinciVisionLLC_Template ☰

VinciVisionLLC_Template.prproj

Name ↑

- Assets
- Sequences
  - Nest
  - Others
  - Output
    - Adset1
      - J9_V1A_H1S1_UGC_MarioB_VV
      - J9_V1B_H1S2_UGC_MarioB_VV
      - J9_V1C_H2S1_UGC_MarioB_VV
      - J9_V1D_H2S2_UGC_MarioB_VV

Example 3

In *Example 3*, the setup is testing **Sales Sequences (S)** along with **Headers (H)**—helpful when the goal is to optimize the core pitch or body of the ad.

A name like **J9_V1A_H1S1** means:

- **J:** *Job 9*
- **V1A:** *Adset 1, Variation: A*
- **H1S1:** *Header 1* and *Sales Sequence 1* being tested

Or **J8_V1B_O1R2** would mean:

- **J:** *Job 8*
- **V1B:** *Adset 1, Variation B*
- **O1R2:** *Opener 1* and *Onramp 2*

It's important to only **include in the name what you're actually testing** in that variation. If you're not testing the outro or the results section, no need to include them in the label. Keep names focused and readable.

How This All Looks in Practice in the Timeline

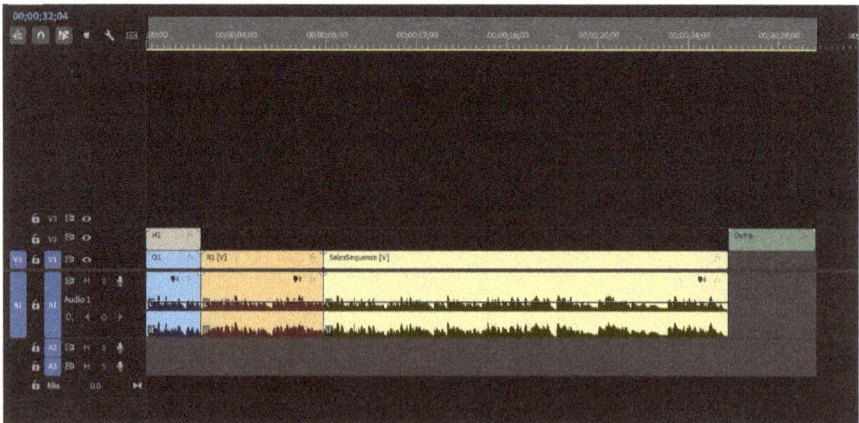

Example A: Full Overview of a Sequence

In the timeline structure shown, the sequence is built from nested segments labeled:

- **H1** = Header 1

- **O1** = Opener 1
- **R1** = Onramp 1
- **SalesSequence**
- **Outro**

Each section is **nested**, you can easily swap out or fix one section across multiple variations. Let's say you realize a specific shot used in R1 is not allowed by the brand and needs to be replaced. Instead of fixing it in 8 different variations, you fix it **once** in the nested R1 sequence — and it updates in every sequence that uses it.

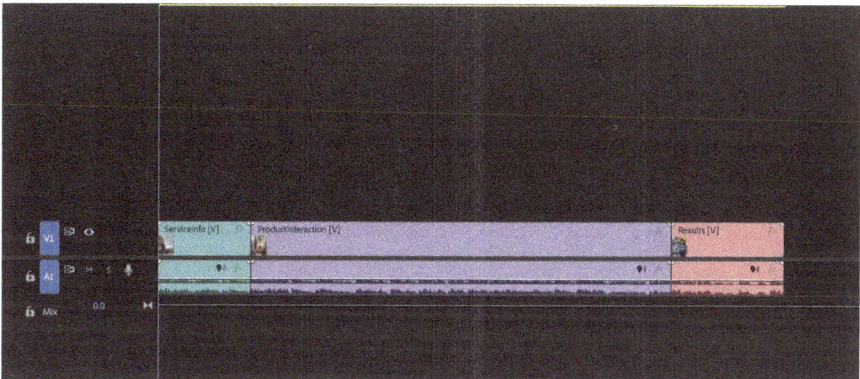

Example B: Inside the Sales Sequence Nest

In this breakdown, the **Sales Sequence** itself is also built with structure:

- **Service Info** – key details about the product or service
- **Product Interaction** – talent using the product
- **Results** – what happened after using it

These are the building blocks of most high-performing UGC ads. By structuring sequences this way and naming them clearly, you set yourself—and your team—up for smarter editing, easier revisions, and better creative analysis.

This kind of clarity not only helps **you** when editing, but also anyone else who jumps into the project — supervisors, agency collaborators, or even your future self revisiting the job a month later.

We'll dive deeper into each structural component (headers, openers, sales sequences, etc.) in future chapters.

4

How Top Agencies Structure Creative Delivery

Using Asana, Monday, and Shared Folders Without Losing Your Mind

Why Organization Matters

Let's get something out of the way: your edits can be amazing, your storytelling flawless — but if your delivery is messy, slow, or confusing, you will absolutely drive your client crazy. And it's not just about being polite. A clean, predictable delivery system makes it easier for clients to test, scale, and request changes. That leads to better performance, smoother collaboration, and more work coming your way.

This chapter is about how to avoid chaos.

The Goal: Organized, Trackable Delivery

Top agencies use project management tools like Asana, Monday, or even ClickUp to keep everything moving. You don't need to use them all. But you do need to understand how they work — because if you're working with a team or client that uses one of these, you'll likely be expected to follow their system.

Here's what these tools are generally used for:

- **Asana:** Great for timeline-based and task-focused teams.
- **Monday:** Highly visual and flexible; ideal for teams that like to customize their workflows.
- **ClickUp:** Combines docs, tasks, timelines, dashboards — a one-stop shop for some teams.

In every case, these tools are used to:

- Assign deliverables
- Track versions and revisions
- Link final files for download or review
- Keep launch plans and deadlines visible for everyone

The VinciVision Delivery System

Let's walk through a real system in action — the one I use every day.

The VinciVision Delivery System

1. The Board Layout

We organize our projects in a simple **Kanban-style board** with five sections:

- **Not Started**: Tasks that are planned but haven't been touched.
- **In Progress**: Editing has started.
- **In Review**: Waiting on feedback.
- **Needs Changes**: Revisions were requested.
- **Ready to Upload**: Fully approved and awaiting final export/upload.

This makes it easy for anyone — editor, project manager, client — to instantly see where things are.

2. The Info That Matters

Each task/card on the board shows:

- **Job Number (J#)** — matches the naming system used in files

43

- **Variations** — how many versions are being made (typically 4)
- **Product & Platform** — which product this video is for, and where it's going (Meta, Amazon, etc.)
- **Aspect Ratio** — 16:9, 9:16, 4:5, etc.
- **AirLink** — our preferred tool for sharing final deliverables (works like Frame.io or Dropbox links)
- **Assignee & Video Editor** — to know who's responsible and who's editing
- **Requester** — usually the creative director, brand lead, or client who asked for the job
- **Due Date & Upload Date** — keeps things accountable

All of this info pulls from the folder structure and naming system we covered in Chapter 3. It's not random — it all works together.

> ☞ **Note:** The only difference is that when we write the job name on the board, we **don't include** the full naming structure like **V1A_H1O1** to avoid overcomplicating things. For example, if we have a job with four variations named J5_V1A_H1O1_UGC_Mari oB_VV_45, J5_V1B_H1O2_UGC_MarioB_VV_45, etc., we'll just write **"J5_UGC_MarioB_VV"** — the same as the job folder name. You'll notice that in the delivery system, there's a separate field for Variations, where we indicate how many variations are included in the job.

The idea here is simple: no confusion, no back-and-forths asking "which one is final?", "where's the square version?", or "who edited this?"

Be a Pro Clients Trust

Even if you're a solo editor, the way you deliver your work speaks volumes about your reliability and professionalism. Top brands don't just want good creative — they want to know it's organized, trackable, and ready to scale. When you can speak their language and match their structure, you immediately stand out.

Whether you're using Asana, Monday, or a Google Sheet, build a simple but clear system like this. Then stick to it.

II

Part II: Execution & Creative Craft

Structure, strategy, and editing techniques. Understand metrics, platform differences, and how to build converting ads — including advanced UGC and the role of tone and clarity.

5

The Metrics That Matter

Note to the Reader:

This is the most technical chapter in the book — and the most important one. It breaks down the metrics that decide whether your edits actually work. Not just "did people like it," but "did it perform."

You'll learn how to interpret CTR, CPC, CVR, CPA, ROAS, engagement rate, and audience insights — and how they all connect. This is the chapter that gives you real creative leverage. The ability to look at ad results, diagnose what went wrong (or right), and actually improve performance.

If you want to be more than an editor — if you want to become a creative strategist who gets results — this is the chapter to master. It might feel heavy the first time through. That's fine. Come back to it as needed. But don't skip it. It's what turns this book from a creative guide into a results-driven playbook.

As a video editor in the paid social ad space, you might wonder how to gauge the success of your creative work. After all, a video can be beautifully crafted, but in advertising, **data tells the story** of whether that video is actually driving business results. In this chapter, we'll demystify *the metrics that matter* for paid social video ads. These key performance indicators (KPIs) will guide your creative decisions and help you speak the language of marketing performance. We'll cover the big ones: **CTR, CPC, CVR, CPA, ROAS,** and **Engagement Rate** – explaining what each means, why it matters, how they connect to each other, and how you can use them to sharpen your edits. We'll also touch on **Audience Insights** (demographic and behavioral data) and finish with **ROI** – what it is, how it differs from ROAS, and why it's important for a creative producer to understand.

By the end of this chapter, you'll not only know the definitions of these metrics, but also how to **interpret them in the context of your video ads** and use them to make decisions that improve performance. Let's dive in.

Click-Through Rate (CTR)

What it is:

Click-Through Rate (CTR) is **the percentage of people who clicked on your ad after seeing it**. In formula terms, if an ad was shown (impressions) and received clicks:

```
CTR = (Clicks / Impressions) × 100.
```

For example, if your video ad was displayed 1,000 times and 50 people clicked on it, the CTR is 5%. CTR essentially measures how enticing your ad is to the audience: it's the **first indicator of whether your creative caught someone's attention and prompted them to take action.**

Why it matters:

CTR is often **the first metric advertisers look at** to judge ad engagement. **A high CTR means a significant portion of viewers found your ad compelling enough to click** – in other words, the ad is resonating with the audience. Platforms like Facebook and Google Ads even reward ads with higher CTR by giving them better relevance/quality scores and more impressions, often at a lower cost, because a **high CTR signals the ad is relevant to users.** Conversely, a **low CTR is a warning sign**: it suggests that something about **the ad** (the visuals, message, or targeting) **isn't clicking with the audience**. It's like a thumb down for your creative – people see it but don't find it interesting enough to engage. Remember, **getting the click is the first hurdle**. If your CTR is low, not many people are entering your funnel to even have a chance to convert.

How it connects to other metrics:

CTR has a direct impact on **CPC** (Cost Per Click) and an indirect effect on downstream metrics like conversions. Think of it this way: if you improve CTR, you'll get more clicks for the same number of impressions. Often, this leads to a lower CPC, since you're essentially becoming more efficient at generating clicks. More clicks also mean more visitors to your landing page, which (all else equal) gives you more chances to convert – but **you'll need a good CVR (Conversion Rate) for those clicks to turn into sales or sign-ups**. In short, **CTR** is the *entry point* of your ad's performance funnel. **It doesn't guarantee success beyond the click** (someone might click but not convert), but without clicks, you have no conversions at all. So, CTR works hand-in-hand with CVR: a strong CTR fills the top of the funnel with potential customers, and a strong CVR turns those potential customers into actual customers.

Example:
In one of my freelance projects for a startup's Facebook ad, the initial CTR was

just 0.5%. This was a red flag – the video was getting impressions but hardly any clicks. To diagnose the issue, I looked at the creative: the opening was slow and the value proposition only became clear 10 seconds in. I re-edited the first 3 seconds to add a bold headline and a striking visual hook right at the start. The result? The CTR jumped to about 1.5% the next week. That threefold increase in CTR meant **3× more people** were visiting the product page from the same number of impressions. This also had a cascade effect: because Facebook saw people engaging more, it started showing the ad more favorably. My cost per click dropped significantly (from ~$1.00 down to around $0.35) without any change in budget, simply because the ad was more efficient at earning clicks. More importantly, those extra clicks eventually led to more conversions down the line. This experience cemented for me how crucial CTR is – it all starts with capturing attention.

Using CTR as a creative decision tool:

As a video editor, you should **monitor the CTR of your ads like a hawk**. It's one of the clearest signals of whether your creative **hook** is effective. If you notice a low CTR on a video ad, consider it constructive feedback. You might need to try a different approach in the edit: for example, make the opening more gripping (since viewers often decide within a second or two whether to keep watching), test a more curiosity-driven thumbnail or title text, or ensure your ad's message matches the audience's interests. On the other hand, **if your CTR is high, that's a sign you've done something right in grabbing attention – you've won the scroll-stopping game.** You might still refine the content after the click, but at least the ad is doing its job in generating interest. In practice, creatives often A/B test multiple versions (also known as "variations") of a video (changing the first few seconds, the Header or Opener) and let CTR be one of the deciding factors for which version to scale up. Remember, **CTR is your early indicator**. It answers the question: *"Are people intrigued enough by my ad to learn more?"* If the answer is no, your creative (or targeting) needs a rethink. If yes, you're on the right track and can focus on what happens after the click.

Cost Per Click (CPC)

What it is:

Cost Per Click (CPC) is exactly what it sounds like – **how much, on average, each click on your ad costs**.

If you spent $100 on an ad campaign and got 200 clicks, your average CPC is $0.50 per click. The formula is:

```
CPC = Total Ad Spend / Number of Clicks.
```

In many advertising platforms (like Facebook or Google), you don't set the CPC directly; instead, it's an outcome of the auction and **how engaging your ad is**. In simple terms, CPC tells you **how expensive it is to get a single person to click your ad**.

Why it matters:

CPC is a crucial metric because **it's tied to your budget and the efficiency of your campaign**. A *lower CPC* means you're getting **more clicks for the same amount of money**, which is generally a good thing – it indicates your ad is either high quality (users click it a lot, driving costs down) or that you're in a less competitive auction. A *higher CPC* means **each visit to your site is costing you more**, which can quickly eat into your return on ad spend. For creative folks, think of CPC as the **financial reflection of your CTR**. If you have a great video that people love to click **(high CTR), the platform often rewards you with a lower CPC**. On the flip side, if your **ad isn't grabbing attention, the platform might charge you more for each scarce click, or you'll only get**

clicks by bidding higher. It's worth noting that some niches naturally have higher CPCs (e.g., ads for expensive software might cost several dollars per click due to competition), but within your control, creative and relevance make a big difference. Ultimately, CPC matters because **it affects how far your budget goes**. If you have a fixed budget and your CPC is cut in half, you can get twice as many people to your site. That could mean the difference between 5 conversions versus 10 conversions, which is huge.

How it connects to other metrics:

CPC is tightly interwoven with CTR and CPA. As mentioned, **higher CTR usually leads to a lower effective CPC because you're generating clicks efficiently**. Mathematically, if you're paying per impression (as many platforms essentially do internally), doubling CTR roughly halves the CPC. Even on bidding models where you pay per click, a high CTR improves quality/relevance scores, which can reduce the amount you need to bid for the same exposure, again lowering CPC. Now, **why does CPC matter down the line?** Because it feeds into **CPA (Cost Per Acquisition)**. If each click costs you $1 and (for example) 1 in 10 clicks converts (10% conversion rate), then on average you pay $10 for one conversion. But if each click costs $2, that same 10% conversion rate yields a $20 CPA – double the cost to acquire a customer. So, **CPC and CVR together determine CPA**. As a creative, you can't directly set the CPC, but by influencing CTR and engagement, you indirectly influence CPC. Also, CPC can be a signal: **if you notice your CPC is way above industry benchmarks or creeping up, it might indicate your ad's relevance is suffering** (possibly due to creative fatigue or wrong audience targeting). In summary, CTR and CPC are two sides of the same coin – **CTR is user behavior, and CPC is the cost outcome**. And both will influence how costly or efficient it is to get conversions (CPA) and revenue (ROAS).

Example:
On a freelance gig, I was editing video ads for a client in the fitness niche. We ran two versions of a video: one had a flashy, upbeat intro, and the other had a more

generic start. Everything else was the same and we split the budget evenly. The ad with the flashy intro not only got a higher CTR, but we observed the CPC was considerably lower – about $0.60 per click versus $1.20 per click for the generic version. This meant our budget went twice as far with the engaging edit. By the end of the test, the engaging video brought in **double the site traffic for the same spend**. This stark difference taught me that editing choices affect CPC indirectly. In this case, the platform (Facebook) likely saw users responding well to one video (higher CTR, higher engagement like shares) and thus delivered it more cheaply. The takeaway for the client was clear: **invest in more creatives like the winning version because they get more value out of each advertising dollar.**

Using CPC as a creative decision tool:

While you're not setting the bids as a creative, keeping an eye on CPC can give you feedback on your ad's performance health. If you notice your **CPC climbing over time for the same ad, it could mean the ad is wearing out** (people have seen it too much and are no longer clicking), **or competitors are outbidding with more engaging ads**. From a creative standpoint, a **rising CPC is often a cue to refresh the creative**. Maybe it's time to edit a new variation of the video or try a new angle to boost that CTR and engagement back up. Additionally, compare CPC across different creatives and platforms. You might find, for instance, your video ad has a nice low CPC on Instagram, but a higher one on TikTok – perhaps the creative needs tweaks to fit the different audience or format.

> **One important note:** a low CPC is good, but it's not everything. Sometimes an extremely catchy ad can drive lots of clicks (low CPC) but those clicks might not convert (perhaps because the ad was clickbait or attracted the wrong audience). So always interpret CPC alongside conversion metrics.

As an editor, aim for ads that **both keep CPC low and drive relevant clicks that convert.** In practical terms, focus on relevance and clarity: an ad that clearly appeals to the right people tends to get cheaper clicks than an ad that's vague or shown to the wrong crowd. In summary, use **CPC as a pulse check. If your CPC is low and CTR high – great**, your video is hitting the mark. **If CPC is high, ask "Is my creative engaging the audience?"** and experiment with changes to find a version that people click willingly (and thus cheaply).

Conversion Rate (CVR)

What it is:

Conversion Rate (CVR) tells you **what percentage of users take the desired action** after initially engaging with your ad. In the context of paid social, CVR often refers to **the percentage of people who clicked your ad that then** *convert* **on the landing page or website**.

It's typically calculated as:

```
(Conversions / Clicks) × 100.
```

For example, if 100 people click your video ad and 5 of them end up purchasing or signing up (whatever your goal is), your CVR is 5%. Sometimes, marketers also talk about conversion rate in terms of **impressions (conversions per impression, effectively CTR × post-click conversion)**, but most often when you see CVR in ad reports, it means *after they clicked, how many converted.* The "conversion" can be any goal you set: a sale, a lead form submission, an app install, etc. So basically, CVR measures **how good your ad (and the follow-up experience) is at getting people to complete the goal**.

Why it matters:

If CTR measures the hook, **CVR measures the follow-through. A high CVR means that the traffic your ad brought in was interested enough to take the next step** – they found what they were looking for, the offer was compelling, and the overall experience (from ad to landing page to action) was smooth. **A low CVR**, on the other hand, indicates a breakdown after the click: **people clicked your ad but then didn't do what you wanted them to do**. This could be due to many factors – perhaps the landing page was slow or confusing, the offer wasn't appealing, or maybe the ad misled them about what to expect. For a creative producer, CVR is important because it **reflects the quality of traffic your ad is driving and how well the ad's message aligns with the outcome**. If you make a video that gets tons of clicks (great CTR) but none of those clickers convert, you haven't actually delivered value to the business. In essence, CVR answers **"Did the clicks turn into customers (or leads)?"** A strong CVR amplifies the impact of a high CTR – it means **not only are people clicking, they are also *buying***. From a business perspective, improving CVR is gold: **if you can get more of the people who click to convert, you get more results without needing more budget.** In terms of ROI and ROAS (which we'll discuss later), CVR is a major driver. As one source succinctly puts it, *"a higher CVR means you are getting more customers or leads from the same amount of traffic, which ultimately translates to a better ROI"*. It's the metric that turns traffic into tangible outcomes.

How it connects to other metrics:

CVR is intimately connected to CPA and ROAS. Let's break it down: Suppose your CPC is fixed; if you double your CVR, your CPA (cost per acquisition) will halve because you're getting twice as many conversions for the same number of clicks/cost.

In formula form:

```
CPA = CPC / CVR (when CVR is expressed as a fraction)
```

Improving CVR directly lowers your cost to acquire each customer. This means even if your video doesn't drastically change CTR or CPC, making sure the people who do click are *primed to convert* is huge. How does a video ad influence CVR? Part of CVR depends on the landing page or product, which might seem outside a video editor's control. However, **the creative sets expectations.** If your video clearly and honestly showcases the product or offer, the people who click are likely genuinely interested – they know what they're getting, so they're more likely to convert. That tends to increase CVR. If the video is vague or just "cool" but not relevant, you might get curious clicks who then bounce (hurting CVR).

Another link: CVR combined with CTR gives you the overall percentage of impressions that convert (sometimes called **Click-through conversion rate** or **impression-to-conversion rate**). For instance, if an ad had a 5% CTR and a 20% CVR (per click), then 1% of impressions turned into conversions. Advertisers often examine these in tandem. And of course, **higher CVR leads to better ROAS/ROI** because you're generating more revenue from the given clicks and spend.

One more connection: CVR can inform **audience targeting** decisions – if certain audience segments have much higher CVR, that insight can guide creative tailoring or budget allocation (more on that in Audience Insights). In short, **CVR is the bridge between *interest* and *action*.** It magnifies the value of CTR and makes the difference between an engaging ad and a high-performing ad.

Example:
I once edited a series of Instagram video ads for an online course. One version of the ad had a very cinematic, mysterious approach – it got a decent CTR because people were intrigued. Another version was more straightforward: it clearly stated in the video what the course offered and even showed a quick testimonial clip.

Interestingly, both versions had similar CTRs (around 1%), but when we looked at conversion rates, the straightforward ad significantly outperformed the mysterious one. The CVR for the straightforward video's traffic was about 15%, whereas the cinematic video's traffic converted at only about 5%. This meant the same number of clicks yielded **three times more sign-ups** with the clearer ad! Why? It seems the straightforward video attracted clicks from people who knew what they were getting and truly were interested – hence high conversion. The cinematic one attracted some curiosity clicks, but possibly those people felt "this isn't what I thought" on the landing page and didn't convert. The lesson I learned: **set the right expectations in your video.** As an editor, you have to balance making an ad enticing (CTR) with making sure it's **qualifying the right viewers** to click. In this case, we pivoted entirely to the style that yielded better CVR, because at the end of the day, conversions were what the client needed.

Using CVR as a creative decision tool:

While conversion rate is influenced by many things (pricing, website user experience, etc.), there are creative levers you can pull to improve it. First, pay attention to the message match: **ensure the content of your video ad aligns tightly with the landing page and the action you want users to take**. If your video touts a 50% off sale, the landing page better be about that sale. **Consistency helps people follow through**. As a video editor, you can also collaborate with copywriters or marketers to include clear calls-to-action (CTAs) in the video that *pre-frame* the conversion ("Sign up for a free trial," "Shop now for 20% off," etc.). This way, **users clicking already have the end goal in mind**.

Another creative aspect is **audience targeting** through content – if your ad's visuals and story speak directly to the *right* audience, the clicks you get will be more likely to convert. For example, if you're advertising a fitness app, a video showing how the app helps *busy moms* might yield sign-ups from busy moms at a higher rate than a generic fitness montage that gets a lot of random clicks. Use CVR data to inform your editing decisions: **if one version of your creative yields higher CVR, analyze why. Did it clarify the value prop better? Was the tone more trustworthy?** You can double down on those elements in future edits. Also, discuss with your team about the **post-click experience**. Sometimes small creative tweaks on the landing page (like using a screenshot from the video, or mirroring language from the ad) can boost CVR – ensuring a seamless journey from ad to action. In summary, **treat CVR as the metric that completes the story started by CTR**. A creative producer who understands **CVR will aim not just to get any click, but to get the *right* click**. By doing so, you'll produce video ads that don't just attract eyeballs – they drive results.

Cost Per Acquisition (CPA)

What it is:

Cost Per Acquisition (CPA) (also known as Cost Per Action, or Cost Per Conversion) is the **average cost to acquire one conversion or customer** from your ads. It's calculated by taking the total amount you spent on a campaign and dividing it by the number of conversions (acquisitions) it generated.

```
CPA = Total Ad Spend / Total Actions
```

If you spent $500 on ads and got 25 purchases, your CPA is $20 per acquisition. Essentially, CPA answers: *"How much did it cost us to get one sale (or lead)?"* This metric is crucial because it directly **ties your ad spend to results**. It's often the bottom-line number advertisers care about: if your **CPA is lower** than the profit you get from a customer, **you're in good shape**; if it's **higher, you're losing money** on each conversion.

Why it matters:

CPA is the clearest **indicator of efficiency and profitability per customer** in advertising. For a creative working on paid social videos, understanding CPA is vital because it's what your client or company ultimately wants to optimize. They might love a beautifully edited video, but **if the CPA is too high** (meaning it's too expensive to get customers from it), **that ad isn't sustainable in a performance-driven campaign**. A **low CPA** means your **ads are doing a great job of turning ad dollars into actual customers** or actions at a cost that the business can afford. A high CPA might signal problems in the funnel – maybe the video isn't attracting the right people, or the offer isn't compelling, or something is off post-click. **Why it matters to you as a creative:** because your creative decisions have a domino effect on CPA. Remember the chain: **your video's CTR and ability to attract qualified clicks affects CPC and CVR, which together determine CPA**. CPA is often the **key metric for campaign success**, especially in performance marketing. If you're freelancing, clients might even pay you or bonus you based on hitting a target CPA. Additionally, when comparing different ads, CPA gives a single metric that encapsulates the impact of all the others: it bakes in CTR, CPC, CVR (since CPA = cost/click ÷ CVR as a fraction, or simply total cost / conversions). **Many advertisers will kill an ad with a high CPA and scale one with a low CPA**. By appreciating this metric, you can reverse-engineer what aspects of your creative might be improved. For example, if an ad has a higher CPA than another, you can dig into why – is it lower CTR, lower CVR, or both? That can guide whether you need to tweak the hook or perhaps the messaging for clarity.

How it connects to other metrics:

CPA is the *culmination* of the metrics we've discussed so far. It connects directly with **CPC and CVR** in particular. A simple approximation: CPA ≈ CPC / CVR. Let's illustrate: if your CPC is $1 and your conversion rate (per click) is 10%, then on average you spend $10 to get one conversion (CPA $10).

CPA formula for diagnosing performance:

```
CPA = CPC / CVR (with CVR expressed as a decimal)
```

This is a **mathematical rearrangement** of the same logic. If:

- CPC = $1 (cost per click)
- CVR = 10% = 0.10 (conversion rate per click)

Then:

- CPA = $1 / 0.10 = $10

This version is useful when you want to **break CPA down into its components** and understand which lever (CPC or CVR) is impacting it.

If either CPC goes up or CVR goes down, CPA rises. Conversely, **lower CPC or higher CVR drives CPA down**. This is why earlier in the chapter we stressed improving CTR (to lower CPC) and improving CVR – because together they make your CPA attractive.

CPA also relates to CTR indirectly: CTR doesn't factor into CPA directly except through affecting CPC and the volume of clicks, but sometimes you'll hear about *Cost per Acquisition per impression* as an overall efficiency (that's essentially CPA combined with CTR). Another connection is with ROAS/ROI. **CPA by itself doesn't include revenue; it's purely a cost metric.** But if you know the value of each conversion (say each sale yields $50 revenue), you can derive ROAS or profit from CPA. For instance, if CPA is $20 and each sale is worth $50, you're doing well; if CPA is $60 for a $50 product, that's a losing game. **CPA is often used alongside average order value or lifetime value to judge if it's acceptable.** Importantly for creatives, **CPA is often the North Star for optimization** in performance campaigns – ads are frequently run on a CPA goal. Some ad platforms even let you bid or optimize for CPA directly (they try to get you conversions at a target CPA). That means if your creative isn't helping hit that target, the platform will favor other ads. In sum, **CPA ties together the efficiency of spend (CPC) with the effectiveness of conversion (CVR).** It tells you in one number **how costly it is to get what you want.** Lowering CPA is usually the primary objective, and all the other metrics are the knobs you can turn to achieve that.

Example:

For a client in the e-commerce space, I produced two video ads for the same product. Ad A had an average CPA of $30, while Ad B's CPA came out around $45. This was puzzling at first because both videos were similar length and showcased the product features, just in different styles. We dug into the funnel metrics and found the difference: Ad A had a slightly higher CTR and significantly better CVR on the site – likely because it set better expectations – whereas Ad B, despite decent CTR, led to a lot of window shoppers who didn't buy, thus fewer conversions. That drove up the CPA for Ad B. By identifying this, I worked on re-editing Ad B to be clearer about the product pricing and offer, effectively targeting more serious buyers. The next iteration of Ad B brought its CPA down to ~$28, now even a bit better than Ad A. This was a win for the client, as we now had two strong ads to scale. The key learning for me was that **an editor's job doesn't end at making something look good – it extends to analyzing why one creative drives cheaper acquisitions than another.** In this case, adding a simple callout in the video about "Free Shipping on Orders Over $50" filtered in more qualified clicks, which improved conversion rate and slashed the CPA.

Using CPA as a creative decision tool:

CPA is often the bottom-line metric you'll be judged on in performance advertising. As a creative, you should e**mbrace this and use it to your advantage.** Keep track of the CPA for each of your video variations when running tests. **If one version consistently yields a lower CPA, study that ad closely.** Was the intro more engaging? Did it highlight the product differently? Did the tone match the audience better? These clues can inform not only tweaks to that ad but also your approach to future projects. For instance, you might discover that lifestyle-oriented footage in the first 5 seconds leads to a lower CPA than pure product shots – perhaps because it attracts a more intent-driven audience. With that insight, you'd incorporate more lifestyle hooks in upcoming edits. Another practice is to **work backwards from CPA goals**. If a client says "We need a CPA of $20 to be profitable," and you know roughly your CPC range, you can estimate the CVR needed and thus tailor the message to drive that kind of intent. This might involve coordination beyond editing – like advising on landing page continuity – but it starts with the creative mindset. Also, consider **testing different CTAs or offers in your videos**. Sometimes a small change like "Get 50% Off – Today Only!" versus "Buy Now" can drastically affect people's motivation to convert, thus altering CPA. Always pair qualitative creative intuition with the quantitative measure of CPA. Over time, you'll develop a sense for what creative strategies tend to yield efficient CPAs in your niche. Importantly, **if CPA is high, don't panic – analyze**. Is it high because clicks are too costly (low CTR issue) or because conversion rate is poor (message/targeting issue)? Diagnose the weak link and address that part of your creative or funnel. This diagnostic approach will make you not just a video editor, but a **problem-solver** who can i**terate toward ads that both look great *and* perform great**. In performance marketing, that's the holy grail.

Return on Ad Spend (ROAS)

What it is:

Return on Ad Spend (ROAS) is a metric that tells you **how much revenue you earned for each dollar spent on advertising**. It's essentially the revenue-to-cost ratio of your ad campaign.

The formula is:

```
ROAS = (Revenue from the campaign) / (Cost of the campaign).
```

Often it's expressed as a ratio or a multiple (e.g., ROAS = 3 means $3 earned per $1 spent) or as a percentage (300% for example). For instance, if you spent $500 on ads and those ads drove $2,000 in sales, your ROAS is 4, or 400%. In plain terms, ROAS answers: *"Are my ads making money, and how efficiently?"* This metric is king for many advertisers because it directly shows the **effectiveness of the ad spend in terms of revenue generated**.

Why it matters:

ROAS is a comprehensive performance indicator. While CPA told you cost per conversion, **ROAS tells you *value* per cost**. **A high ROAS means your campaigns are bringing in a lot more money than you're spending** – that's usually a very good sign! A ROAS **above 1** (or above 100%) indicates **you're earning more than you spend**, whereas a ROAS **below 1** means **you're spending more than you're earning (not sustainable long-term)**. For a creative producer, understanding ROAS is crucial because it encapsulates the end-to-end success of your ad creative in driving profitable outcomes. An ad with a great CTR and low CPA might still have a poor ROAS if, say, it's promoting a low-priced product without enough revenue to cover costs. Conversely, an ad with a moderate CTR but for a high-value product could have an excellent ROAS. **ROAS is what the business really cares about at the end of the day** – *"If I give you $1 for ads, how many dollars do I get back?"*. By keeping ROAS in focus, you as a creative can align your goals with the business goals. For example, if you know a certain product has high margins (thus contributing strongly to ROAS), you might prioritize making winning creatives for that product. Additionally, ROAS can influence budget decisions: **campaigns with high ROAS get more budget, and low ROAS campaigns get cut.** Your creative performance directly affects that.

How it connects to other metrics:

ROAS is linked to all the metrics we've discussed because it's the final result of the chain. Let's connect the dots:

> CTR affects how many people enter the funnel > CPC affects how much you pay for those people > CVR affects how many of those clicks convert > average order value (which we haven't deeply discussed but is the revenue per conversion) combined with number of conversions gives total revenue. **All of that rolls up into ROAS.**

1. The Core ROAS Formula
At its most basic:

```
ROAS = Revenue / Cost
```

This is the one that everyone uses in reporting. If you made $1,000 from $250 in ad spend → your ROAS is 4.0 (i.e., $4 made for every $1 spent).

2. Expanded ROAS Formula
The point of expanding it is to **diagnose or understand what's impacting ROAS** behind the scenes.

First, split Revenue:

```
Revenue = Number of Conversions × Revenue per Conversion (AOV)
```

And recall:

```
Conversions = Clicks × CVR
```

So:

```
Revenue = (Clicks × CVR) × AOV
```

Now, split **Cost**:

```
Cost = Clicks × CPC
```

3. Now Plug Both into the ROAS Formula

```
ROAS = Revenue / Cost

ROAS = [(Clicks × CVR) × AOV] / (Clicks × CPC)
```

Now cancel out the **Clicks** (on both top and bottom of the fraction):

```
ROAS = (AOV × CVR) / CPC
```

What Does This Mean?

It tells us **what directly affects ROAS**:

- **AOV** (Average Order Value) → if customers spend more, ROAS goes up.
- **CVR** (Conversion Rate) → if more people convert after clicking, ROAS goes up.
- **CPC** (Cost Per Click) → if it costs you less per click, ROAS goes up.

But Isn't That Complicated?
Yes — **but only if you try to memorize it like separate formulas**.

The truth is:

All of these formulas are just different expressions of the same system.

They let you look at the problem from **different angles**:

- ROAS = Revenue / Cost → Big picture
- ROAS = (AOV × CVR) / CPC → Performance-focused breakdown
- CPA = CPC / CVR → Cost-focused lens
- CTR doesn't show up directly, but it **affects CPC**, which affects CPA and ROAS

Think of these formulas like zoom lenses:

- **ROAS = Revenue / Cost** → zoomed-out view
- **ROAS = (AOV × CVR) / CPC** → zoomed-in mechanics
- **CTR → CPC → CPA → ROAS** → the *domino effect*

It's not about using *all* the formulas at once — it's about knowing **which version helps you troubleshoot or optimize.**

Higher CVR improves ROAS, higher average revenue (e.g., selling higher priced items or increasing cart value) **improves ROAS, and a lower CPC improves ROAS.** If you boost CTR and keep quality of clicks high, you lower CPC and often can increase conversions, which helps ROAS. Engagement rate too can indirectly help ROAS by lowering costs or increasing reach. One thing to note is that ROAS doesn't explicitly account for CTR, except as it influences those other factors. **You can even think of CTR as the *engine* that drives the volume which, combined with conversion efficiency and value, yields ROAS.** It's all interconnected. For example, imagine you have an ad with modest CTR, but those who click buy expensive products at a high rate – you can still have great ROAS. Or an ad with high CTR but low-value conversions could have mediocre ROAS.

All metrics contribute: CTR gets people in > CVR and AOV (average order value) convert value out of them > CPC/CPA determines cost efficiency > ROAS tells you the net effect.

Also, ROAS is closely related to ROI (Return on Investment), which we will cover shortly – **ROAS is specifically about *ad spend* return, whereas ROI includes all costs**. Many advertisers target a specific ROAS (like "we need at least a 3x ROAS to be profitable given our margins"). **If your campaign's ROAS is below target, it signals something's off in the mix of CTR, CVR, targeting, or creative messaging that needs addressing.**

Example:
One of my e-commerce clients was primarily concerned with ROAS for their campaigns. We ran a video ad that had a CPA of $25 and each conversion (a sale) was worth $100 in revenue on average. That's a 4x ROAS (400%); in other words, $100 revenue / $25 cost = 4. They were quite happy with this. Later, we tried a different creative approach – a more humorous video. That ad got a ton of engagement and clicks (CTR went up, CPC went down), and the CPA even dropped to ~$20, which looked great. However, interestingly the ROAS on that ad was only about 2.5x (250%). Why? It turned out that while we got more purchases, the average order value was lower – people were buying the cheaper items on the site, perhaps attracted by the humor but not necessarily the high-ticket products. Total revenue didn't rise proportionally to the increase in conversions. This was a fascinating learning moment: **a creative change can shift not just *how many* convert, but *who* converts and *what* they buy.** We adjusted the targeting and messaging to emphasize a premium product in the humorous ad, and then we saw the ROAS climb up to roughly 3.5x as higher-value orders came in. For me, this underscored that ROAS is the ultimate balancing act metric – **it captures both the cost efficiency (CPA) and the revenue side.** It also taught me that as a creative, I should **be aware of the value** of what I'm promoting, **not just the volume**. If a certain edit appeals to bargain-hunters, it might lower ROAS if the business relies on big sales. In this case, by tweaking the

creative to highlight the premium product's benefits (while keeping the fun tone), we aligned the ad with higher-value conversions and improved ROAS.

Using ROAS as a creative decision tool:

ROAS is a metric that might seem a bit removed from the editing process, but it's actually where creative and business outcomes meet. When you review performance, always include ROAS in your analysis. **If two different videos are being tested and one has a higher ROAS, dig in to understand why.** Is it attracting more valuable customers? Is it aligned with a more profitable product line? You can use those insights to guide future creative strategy. For example, you might realize that a particular style of video ad leads to customers who not only convert, but also spend more. Perhaps the storytelling in the video appeals to a customer segment that has a higher lifetime value. Those are exactly the kinds of creatives you want to produce more of. On the flip side, if an ad has a low ROAS, figure out if it's because of low conversions or low revenue per conversion (or both). If conversions are low (and thus revenue is low), you know to optimize CTR/CVR. If conversions are fine but revenue per customer is low, maybe the creative could encourage larger purchases – for instance, by showcasing product bundles or premium options in the video. This is a more advanced tactic, but top-performing creatives often collaborate with marketing on how to **increase AOV (Average Order Value)** via messaging ("Buy 2 get 1 free" or "Free shipping on orders over $50" in the ad can drive bigger cart sizes, boosting ROAS). Another way creatives influence ROAS is by **targeting the right audience through content**, similar to CVR. **If your video resonates with a high-value demographic, you'll likely see better ROAS**. Use any audience insights (coming up next section) to tailor content to those who bring the most value. In summary, **think of ROAS as the final exam score for your ad**. All the little quizzes (CTR, CPC, CVR, CPA) culminate in this grade. As a creative, you want to eventually produce ads that ace that final exam. Keeping an eye on ROAS during testing will help you prioritize which creative directions are truly paying off and deserve scaling. It encourages you to **consider not just "will** this get clicks?" but also **"will this drive revenue effectively?"** – a hallmark of a high-performing creative in paid social.

Engagement Rate

What it is:

Engagement Rate **measures how much people interact with your ad beyond just viewing it.** It typically **includes actions like reactions (likes, loves, etc.), comments, shares, saves, or any clicks** (including not just link clicks but also clicks to play, expand, etc. depending on platform) – basically **any form of engagement divided by the total impressions**.

The formula can be expressed as

```
Engagement Rate = (Total Engagements / Total Impressions) × 100.
```

For example, if your video ad was shown 1,000 times and it got 50 total engagements (say 30 likes, 10 comments, 10 shares), the engagement rate would be 5%. It's a broader metric than CTR, which counts only clicks on the link/call-to-action. **Engagement rate captures the *social interactions* with your ad content.**

Why it matters:

In the world of social media advertising, engagement can be a sign of resonance and virality. **A high engagement rate means people aren't just seeing your ad – they're reacting to it, talking about it, and even endorsing it by sharing.** This is a strong indicator that your creative is striking a chord emotionally or intellectually. For a video editor, that's a point of pride and an important consideration: an engaging ad often means the content is

compelling, relatable, or noteworthy. But beyond pride, there's a practical reason engagement matters: **many social platforms' algorithms reward ads that get good engagement by showing them more and sometimes at lower cost.** Think of Facebook's relevance score (now part of quality ranking) – if your ad gets lots of positive interactions (and not hidden or reported), Facebook perceives it as relevant and gives you a break on reach and CPC. That can amplify your campaign's effectiveness. Additionally, **each share can lead to additional free impressions (organic reach) that you're not paying for, which can indirectly boost your ROAS.** Engagements like comments can also provide feedback – if people are asking questions or tagging friends, you're generating conversation and interest that might lead to more conversions down the line. However, note that not all campaigns prioritize engagement – if you're purely going for conversions, sometimes engagement is a nice-to-have. But often, especially in paid social, **engagement rate is considered alongside CTR to gauge how well the ad creative is connecting with the audience on a social level.** It's also a good metric for **upper-funnel goals** like brand awareness and community building. For an editor, if lots of people are engaging, it usually means your video evoked a response – and getting any response is generally better than being ignored.

How it connects to other metrics:

Engagement Rate and CTR are related but not the same. Some people might watch your video and hit the Like button or leave a comment without clicking through to your website. You could have a high engagement rate but a moderate CTR. Ideally, though, good engagement supports a healthy CTR. If people are moved enough to comment or share, they may also be likely to click through or at least remember your brand. Importantly, as mentioned, **engagement can influence CPC**. A well-engaged ad often enjoys a lower CPC because the platform sees it as valuable user content. This means indirectly, a higher engagement rate can lead to a lower CPA and better ROAS since you're effectively getting cheaper traffic.

Another connection: engagement feedback can guide optimization. For example, reading comments might reveal audience sentiments – maybe people love a certain feature shown in the video (you might emphasize that more), or they are confused about something (you clarify it). Shares indicate word-of-mouth potential; if one of your videos is getting shared a lot, it's a creative home run in terms of virality. Engagement Rate doesn't directly measure conversions, but it contributes to the **holistic picture of ad performance**.

Sometimes, especially with video ads, you might also look at **Video Engagement metrics** like watch time, view-through rate, etc., which are related concepts (how much of the video people watch is a form of engagement). All these help you understand if your creative is holding attention.

Finally, engagement is tied to **Audience Insights**: different demographics might engage differently. Maybe younger audiences share more, older audiences click more – knowing this can inform how you craft content for each group. In summary, **engagement rate is a bit of a "quality and interest" barometer for your creative.** It complements the direct-response metrics (CTR, CVR) by adding a layer of audience sentiment and platform favorability.

Example:
I created a playful TikTok-style ad for a mobile app, using trending music and a bit of humor. The engagement rate went through the roof – people were loving it, commenting with laughing emojis, and sharing the ad. Specifically, the ad got about a 12% engagement rate (lots of likes and shares out of the impressions) which is quite high. Interestingly, the CTR was good but not as astronomical (around 1.2%). However, what we noticed was the campaign's overall costs were very efficient: the high engagement seemed to drive the CPM (cost per thousand impressions) down because the platform was giving us love for the great interaction. In effect, even those who didn't click were boosting the ad by engaging with it, which allowed others to see it for cheaper, leading to more clicks in volume. The outcome was that this ad had one of the lowest CPCs and

CPAs in the account, even though a lot of its success was coming from people simply enjoying and sharing it. One comment even said, "I wasn't even interested in the app, but this ad is hilarious!" – ironically, that comment itself gave the ad more reach, possibly finding someone who *was* interested. This example showed me that **engagement has real value**: it's not just vanity. An entertaining, well-received ad can create a positive feedback loop with the platform's algorithm. Plus, the client was happy to see their brand getting positive public reactions, which is a nice side benefit in terms of branding.

Using Engagement Rate as a creative decision tool:

When you craft social video content, think about elements that **encourage engagement.** This could be as simple as **a relatable joke, a bold statement, or even a question in your ad copy that invites people to comment**. As an editor, you can't force people to like or share, but you can include **engagement cues**. For instance, showing user-generated content or testimonials might prompt viewers to tag friends ("Hey, this is so us!"), or editing in a way that's native to the platform's style (like a TikTok meme format) can tap into existing engagement behavior. Keep an eye on the engagement metrics for your ads. If one video has an engagement rate of 10% and another similar one is 2%, ask why. Maybe the 10% one touched on a cultural trend or had a more provocative opening. Those are clues to what *moves* your audience. **Incorporate those findings into future edits**. Also, consider engagement when defining success with clients: if a campaign's main goal is awareness, engagement rate might even be a primary KPI. You would then optimize your editing for shareability – perhaps making the content funny, inspirational, or discussion-worthy.

> **One caution:** ensure that the engagement you attract is positive or at least on-topic. Sometimes controversial or off-tone content can get engagement (people arguing in comments, etc.), but that might not help the brand or conversion goals.

Aim for relevant engagement. When you read through comments, take them as free focus group feedback on your creative. People often don't filter their opinions in comment sections, for better or worse. You might learn that a certain scene in your video really stood out (positive) or that something was unclear (negative). Use that to refine your work. In sum, **engagement rate is like the applause (or boos) from the audience – it tells you how the crowd reacted.** As a creative performer, you want applause. It feels good, yes, but it also signals that the performance (ad) is connecting, which ultimately can drive better results across the board.

How the Metrics Interconnect: The Big Picture

By now we've defined each metric individually – but in practice, you should see them as parts of one system. Changes in one metric often ripple through to others. As a creative, thinking holistically will make you truly effective. Here's a simple breakdown of how these metrics feed into each other and ultimately into success:

- **CTR influences CPC and traffic volume:** A higher CTR means more clicks from the same impressions. This often lowers your CPC (since you get more clicks per dollar spent) and drives more visitors to your site, which gives you more chances to convert.

- **CPC and CVR determine CPA:** If CPC is low and CVR is high, your Cost Per Acquisition will be low – *that's the dream*. If either goes in the wrong direction (CPC up or CVR down), CPA will climb. For example, doubling your CVR can halve your CPA, improving efficiency dramatically.

75

- **CPA and Average Value determine ROAS:** If you know your CPA and how much a conversion is worth, you effectively know your ROAS. Lowering CPA improves ROAS (since you spend less for each $ of revenue). Also, *increasing* the revenue per conversion (through upsells, etc.) will boost ROAS. ROAS is like the final report card that says "all these metrics together produced X return."

- **Engagement Rate underpins CTR and lowers CPC:** Good engagement often correlates with strong CTR – an ad that people like tends to also be one they click. Even for those who don't click, their likes and shares can expand your reach or signal algorithms to give you cheaper impressions, effectively lowering CPC and helping the whole funnel.

To visualize the interplay, consider this flow:

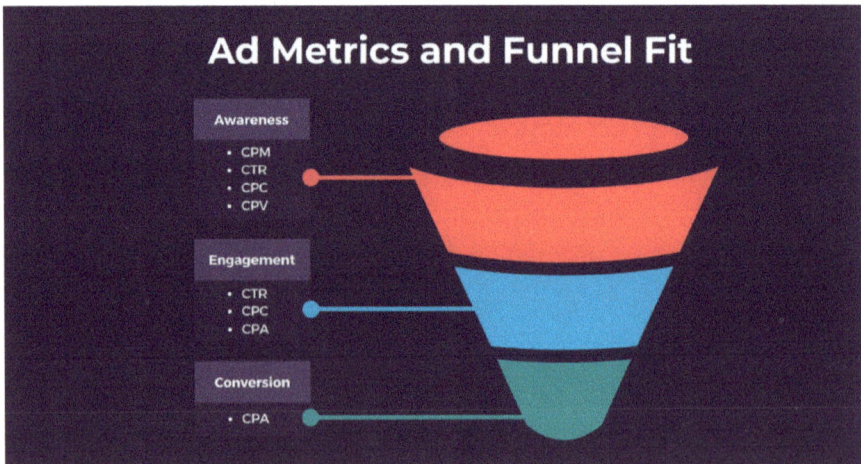

Pathlabs (2024). "A Guide to Ad Metrics."

*Interplay of key metrics in the funnel. A **compelling video ad** (good content) -> leads to a **high CTR** (lots of clicks from impressions) -> which in turn tends to **lower CPC** (cost per click drops as platforms reward relevance). With more clicks coming in cheaply, the focus shifts to converting them: a smooth landing page and aligned messaging yield a **strong CVR** (high percent of clickers convert). Together, low CPC and high CVR give a **low CPA** (each conversion doesn't cost much). If each conversion also brings in good revenue, you achieve a **high ROAS**. Throughout, a **high Engagement Rate** (comments, shares) boosts reach and signals quality, indirectly improving these metrics.*

In short, **all the metrics work together**. You can't look at one in isolation for too long. As a creative, you might start by chasing a better CTR with a flashy edit, but you must also watch what happens to CVR and CPA as a result. A change that improves CTR but hurts CVR could end up neutral (or negative) for ROAS. Conversely, a tweak that maintains CTR but filters out unqualified clicks (lowering volume but boosting CVR) might improve ROAS. It's a balancing act – like spinning plates, you have to keep an eye on all of them, and nudge the right ones at the right time. The beauty is that a well-crafted ad often improves multiple metrics together (e.g., more engaging content can simultaneously raise CTR, engagement, and CVR by attracting the right people and pleasing the algorithm).

Whenever you evaluate an ad's performance, step through this chain of logic: ***"Is my CTR healthy? How's that affecting CPC? Given CPC, is the CVR high enough to yield a good CPA? And does that CPA make sense for our ROAS goals?"*** By answering these, you pinpoint where a creative intervention is needed. Maybe it's up top (the ad isn't enticing -> fix CTR), or maybe the ad is fine but attracting the wrong audience (great CTR, poor CVR -> perhaps tweak messaging targeting). This systemic view will make you a strategic video editor, not just a technical one.

Audience Insights: Shaping Creative with Data

Up to now, we've talked about metrics that directly measure ad performance. **Audience Insights** is a slightly different beast – it's about understanding *who* **your viewers are and** *how* **different groups respond**, so you can craft better creatives for them. Most ad platforms (Facebook, Instagram, YouTube, etc.) provide data on the demographics and behaviors of people engaging with your ads. For example, you might see that women aged 25-34 have a 2× higher CTR on your video ad than men of the same age, or that viewers in certain regions are more likely to watch your video to completion.

Why it matters to creatives:

Knowing your audience data helps you tailor your video editing and storytelling to resonate with the people most likely to convert. It also can inspire new creative angles. Let's say your audience insights show that a certain demographic (e.g. college students) is heavily engaging with your ad – you might double down and create a variant of the video that uses language, music, or references that appeal to that group. Or if you find a key segment isn't engaging at all, you might experiment with a different approach in hopes of reaching them. Audience insights often include: demographics (age, gender, location), interests/affinities, and sometimes behaviors (e.g., "people who often shop online" or "movie lovers"). These can be used to inform creative decisions. For instance, if your product appeals across ages but you see younger audiences respond better to fast, edgy edits, you might create one style of video for them and a different style (slower, more informative) for older audiences.

Demographic tailoring:

Different demographics respond to different creative cues. A Gen-Z audience might appreciate quick cuts, trending audio, and subtitles (since they often watch without sound). An older audience might respond better to a clear narrative and a personable voiceover. If your audience data tells you that 80% of your converters are 45+ years old, leaning into a style that suits them can improve performance. This doesn't mean stereotype; it means test variations that speak their language. Platforms like Facebook Audience Insights or Google Analytics can also tell you about your audience's interests. For example, you might find that people converting on your fitness product ad also have an interest in nutrition and healthy recipes. That's a cue that maybe your next video ad could mention diet tips or show a food scene to grab attention.

Behavioral and contextual data:

Sometimes you learn that most of your viewers are on mobile vs desktop, or that they tend to watch at certain times of day. Creatively, if you know most of your audience is on mobile, you'll ensure your captions are readable on small screens, and perhaps you'll favor vertical formats. If you know evenings are prime time, you might incorporate that knowledge by referencing "after a long day at work..." in your copy, making the ad feel timely.

Audience feedback:

As mentioned in the Engagement Rate section, comments and interactions can also provide qualitative insight. Maybe you notice a trend in comments like "I wish they showed X feature" or questions about the product. That's audience insight telling you what to include in your next edit. Or you might see comments like "Perfect for busy moms!" – aha, maybe busy moms are a key audience; consider making an ad explicitly targeting that persona (e.g., footage of a mom using the product, or adding text like "Calling all busy moms..." to hook that group).

Creative segmentation:

In some cases, you'll actually **produce different creatives for different audience segments** (a practice known as segmentation). For example, a financial service might run one ad featuring a young professional storyline and another ad featuring a nearing-retirement storyline, each to resonate with those respective age groups. The metrics might tell you each performs best with its intended demo. This is a powerful strategy: rather than one-size-fits-all, you're customizing. As an editor, this could mean re-cutting footage to create several versions, each with slight tweaks (different text overlay, different ending scene, etc.) that align with what appeals to, say, New York urbanites vs. Midwest suburbanites if your data shows divergence in those audiences. In practice, use audience insights in a cycle:

```
data -> creative tweak -> test -> data -> ....
```

If data shows a certain audience is valuable, think of what creative elements would speak to them (cultural references, language tone, imagery). Implement and test. Then see if the metrics (CTR, engagement, CVR) improve for that segment. A concrete example: suppose you're advertising a travel app, and you see that 70% of your converters are 25-34 and love adventure travel. **You might cut a version of your video that emphasizes hiking, backpacking, adrenaline shots – things that resonate with adventurers** – and use it when targeting that group. For an older crowd maybe more interested in leisure travel, you'd show relaxing beach or tour scenes in another video. **Tailoring like this often boosts performance because it feels personalized to the viewer**. As one guide noted, segmenting your audience and tailoring content for each can significantly improve engagement and conversion

Audience saturation and fatigue:

Insights can tell you **if you're hitting the same people too often (frequency) and if some people have seen the ad multiple times.** Creatively, if a key audience has seen your ad a lot, you may need a fresh edit to keep them engaged (people tune out if they see the same thing over and over). Insights might prompt you: "Time to refresh creative for women 25-34, our frequency is 5+ for them."

In essence, audience insights turn vague "know your audience" advice into concrete data points you can act on. They help you move from making an ad that *you* think is good, to making an ad that data suggests *the audience* will think is good. The result is often a better alignment with viewer preferences and higher performance.

Beyond ROAS, the Bigger Picture

ROI

We've talked a lot about **ROAS – return on ad spend**. Now let's discuss **ROI (Return on Investment)** and why it's slightly different, and important, especially for a creative producer working closely with marketing goals.

ROI (Return on Investment) in the context of marketing measures **the overall profitability of an investment,** not just the advertising cost.

It's typically defined as:

```
ROI = (Profit from the campaign / Cost of the campaign) × 100%.
```

In other words, **ROI considers *all* costs and returns, not just ad spend.** For a campaign, the "investment" might include the ad spend, but also production costs, your editing fees, any overhead, cost of goods, etc., and the "return" would be the net profit (revenue minus all those costs). If you spent $10,000 on a campaign in total (ads + video production, etc.) and it generated $15,000 in profit after cost of goods, your ROI is (($15k - $10k) / $10k) × 100% = 50% ROI.

ROAS vs ROI:

It's easy to confuse the two. Here's the key difference: **ROAS looks at revenue generated per ad dollar** (it's like a gross measure focusing only on ad spend), while **ROI looks at net profit per total dollar spent** (a broader, net profitability measure).

Another way to say it:

- **ROAS** asks, **"Did my ads bring in more revenue than they cost?"**
- **ROI** asks, **"After all expenses, did the campaign make money, and how much?"**.

For example, imagine an ad campaign where you spent $1,000 on ads and got $5,000 in sales. At first glance, ROAS is 5x (500%) – great. But suppose the product cost to produce was $3,000 and you also paid $500 to a videographer (that's part of investment cost). Your total costs = $1,000 (ads) + $500 (production) + $3,000 (product costs) = $4,500. Revenue = $5,000. Profit = $500. ROI = ($500 profit / $4,500 cost) ≈ 11%. That's much lower – it tells a more cautious story. If those costs were higher than revenue, ROI would be negative, even if ROAS was above 1. This scenario illustrates why a campaign can have a strong ROAS but still not be truly profitable once all factors are accounted.

Why ROI matters to a creative producer:

You might think ROI is more of a finance thing, but it's useful for you to understand because it **contextualizes your work in the larger business picture**. For instance, if you create an ad that has a 300% ROAS, that sounds great, but if the margins are slim, the company's actual profit might be marginal. Knowing this, you could make creative decisions to improve profitability – maybe emphasizing a higher-margin product in the ad or encouraging larger basket sizes. Also, understanding ROI helps you speak the language of stakeholders who care about dollars in vs. dollars out at the highest level. If you propose a new video idea that requires a higher production budget, you better be prepared to argue how it could improve ROI (not just ROAS) – perhaps by significantly lifting conversion rate or tapping a new lucrative audience. Additionally, ROI might factor in *long-term* returns, not just immediate sales. For example, an ad campaign might lose money on first purchase but acquire customers who repeat-buy and become profitable later. That initial campaign's ROAS might be below 1, but the overall ROI considering customer lifetime value could be positive. As a creative, being aware of this can help you frame campaigns appropriately (maybe the goal of a video is to get people in the door knowing future sales will come).

ROI in creative testing:

Suppose you have two ads: Ad X has a ROAS of 4 but was very expensive to produce (maybe you hired actors and did a big shoot), Ad Y has a ROAS of 3 but was dirt cheap to produce (stock footage and quick editing). Which is better? ROAS alone says Ad X is better at generating revenue per ad dollar. But if Ad X's production cost eats up the difference, Ad Y might actually have a higher ROI for the business. This doesn't mean you avoid high production – it means you weigh the investment. If Ad X's concept is scalable and can be used across channels, maybe the investment is justified. But ROI thinking prevents you from getting tunnel vision on ad metrics and forgetting costs. It encourages efficiency in production as well as in ad spend. To distinguish succinctly:

- **ROAS** – **"Revenue-centric."** *Did we earn more than we spent on ads?* For every ad dollar, how many dollars back? (E.g., 3x or 300% means $3 back per $1 ad spend.)

- **ROI** – **"Profit-centric."** *Considering all costs, was it profitable?* For every total dollar spent (ads + production + etc.), how many dollars back in profit? This can be a percentage. (E.g., 50% ROI means the profits were 50% of the investment, i.e., 1.5× return net.)

Both are important, but ROI is the bottom line. A creative producer who understands ROI can better argue for resources ("We should invest an extra $5k in creative because it will likely improve conversion and thus ROI") and also take responsibility for the outcomes ("Our fancy shoot was cool, but it didn't increase conversion enough to cover its cost, hurting ROI – let's pivot strategy").

CAC & LTV

As you dive deeper into creative strategy and client collaboration, you'll start to hear metrics like **CAC** (Customer Acquisition Cost) and **LTV** (Lifetime Value) come up — especially from founders, media buyers, and brand strategists. While these aren't metrics you'll see inside Ads Manager, they're essential for understanding the bigger picture of how your creative impacts a business.

Customer Acquisition Cost (CAC) tells you the average amount spent to acquire a new customer. In many cases, it's similar to CPA — but CAC typically takes into account *all* sales and marketing costs, not just ad spend. This includes everything from production, to email flows, to influencer costs. If you're producing ads for a founder-led brand, they'll often have a CAC target in mind — like, "We need to acquire customers for under $40." Creative that drives a low CPA helps keep CAC low and improves the brand's overall profitability.

Lifetime Value (LTV), on the other hand, tells you how much revenue the average customer brings over time. If a customer signs up today and spends $50 every few months for two years, their LTV might be $300+. For high-LTV businesses (like supplements, SaaS, or subscription brands), they can afford a higher CAC — meaning they're more willing to invest in creative testing, lead-gen hooks, or longer-form ads if it brings in quality customers.

If CAC tells you what it *costs* to bring someone in, LTV tells you what they're *worth*. Together, they inform the ultimate question: "Can we scale this campaign profitably?"

From a creative standpoint, this matters more than you think. Some ads might drive a high CPA, but if they attract high-LTV customers, they're still worth scaling. Others might drive cheap traffic that never buys again. As you grow, start asking clients about CAC and LTV — it shows you understand the full funnel, and it'll help you build creative that's optimized not just for clicks, but for sustainable growth.

Conclusion

As a video editor aiming to be a high-performing creative, **mastering these metrics** – CTR, CPC, CVR, CPA, ROAS, engagement rate, and using audience insights – **is part of your job description.** They might have seemed like scary marketing jargon at first, but now you know they're logical, interconnected measures that you can **leverage to make decisions**. When you **plan and edit videos with these metrics in mind**, you'll find yourself creating not just *art*, but *art that sells*. And that is what turns a video editor into a truly indispensable creative strategist in the paid social ad space. **Keep this chapter as a reference**, keep testing and learning from the numbers, and you'll continue to hook the audience, edit for impact, and convert viewers into customers – living up to the mantra *Hook. Edit. Convert.* every step of the way.

6

Reverse-Engineering a Winning Ad (Using Data to Fuel Creative Experiments)

Once you've created an ad that performs well — or inherited one from another editor — your next job isn't to celebrate and move on. It's to *pull it apart.* Dissect it. Learn from it. And then use what you learn to iterate, test, and scale.

Top performers don't just happen. They're often the result of subtle patterns and repeatable ingredients. The smart editor knows how to read those patterns and build new ads from them.

The Key Metrics to Analyze (Quick Refresher)

If you're new to reading performance data, here's a short breakdown of the most important metrics you'll use to dissect your ad. (We cover these in detail in Chapter 5, but here's what you need to remember for this phase.)

Ad Metrics Breakdown

	Metric	What It Means & Why It Matters
1	Thumbstop Rate	% of people who stop scrolling within the first 3 seconds. This tells you if your opener+header combo is grabbing attention.
2	Hook Rate	Similar to Thumbstop, often used interchangeably. Measures engagement in the very beginning of the ad.
3	Retention Drop-Off	Where viewers begin to exit the video. Tells you exactly what section might be too slow, unclear, or irrelevant.
4	Midpoint Retention	% of viewers still watching halfway through. Helps determine if your Sales Sequence is keeping people engaged.
5	CTR (Click-Through Rate)	% of people who clicked after watching. Reminder: This tells you if your ad inspired action.
6	Engagement Rate	Likes, shares, and comments. While not always tied to conversion, this shows interest and can amplify your reach.
7	CVR (Conversion Rate)	% of people who took the desired action after clicking. Reminder: This is about effectiveness beyond the ad itself.

Where Each Metric Lives in the Ad

Use this cheat sheet when breaking down your performance:

Ad Structure Breakdown

	Ad Section	Timeframe	Key Metric(s)	Purpose
1	Header + Opener	0–3s	Thumbstop Rate, Hook Rate	Visual + copy combo that stops the scroll
2	Onramp	3–8s	Hold Rate, Retention Drop-Off	Sets up personal context or pain point
3	Sales Sequence	8–20s+	Midpoint Retention, CTR, Engagement Rate	Delivers product/service info, demos, results
4	Outro/CTA	Final 3–5s	CTR, CVR	Directs the viewer to act (buy, click, sign up)

Header + Opener Combinations

In performance editing, you don't test headers and openers separately. They always appear *together* in the first 3 seconds, and should be analyzed as a pair.

Let's say you run 4 versions in an adset:

- **V1A_H1O1**
- **V1B_H1O2**
- **V1C_H2O1**
- **V1D_H2O2**

If **V1A_H1O1** has the best Thumbstop Rate, then H1O1 is your top-performing combination. That's your control — a baseline you can pair with new onramps, new tones, or different Sales Sequences to continue scaling.

How to Dissect the Ad

Break it into 4 main parts:

1. **Header + Opener (0−3s)** – One unit. This is your hook combo. Make it visual, bold, and emotional. If it's not performing, test both elements *together*.
2. **Onramp (3−8s)** – Builds personal connection or sets the problem. This is where engagement can drop fast if you lose the viewer's trust or interest.
3. **Sales Sequence (8−20s+)** – The core content. Features, benefits, transformations, or testimonials. The data here tells you how compelling your pitch is.
4. **Outro/CTA (Final seconds)** – Drives clicks or conversions. Make it

strong, clear, and brand-aligned.

Testing Based on Performance Signals

Testing Based On Performance Signals			
	Performance Signal	Creative Insight	What to Test
1	Low Thumbstop	Weak header+opener combo	Test new combinations of copy and visuals
2	Onramp Drop-Off	Weak or confusing story	Try tighter setups or different tones
3	Sales Sequence Drop-Off	Too long, too static	Cut fluff, reorder proof/results earlier
4	Weak Outro/CTA	Soft call to action	Push urgency or simplify next step

Let's say:

- **Thumbstop Rate is high** → opener and header combo is strong
- **Drop-off at 5s–8s** → onramp likely needs clarity or better storytelling
- **Steady decline after 12s** → Sales Sequence could use better pacing or visuals

Your next variations could look like:

- Keep H1O1 from top-performer
- Test 3 onramps: one personal story, one statistic-led, one emotional hook
- Try new order in the Sales Sequence: results first, then demo

The Experiment-First Mindset

Don't treat a great ad like a finished product — treat it like a data-driven blueprint.

Ask:

- "What made this work?"
- "What's the *next smart version* I can test?"
- "What can I re-use and recombine in a new way?"

Every time you analyze a top performer, you're not just learning — **you're engineering your next winner.**

7

High-Converting Ads – Structure, Style, and Substance

Ever wonder why some ads instantly grab your attention while others get scrolled past without a glance? In the fast-paced world of social media, you have only a split second to hook a viewer. That's why mastering the structure, style, and substance of your ad is crucial. In this chapter, we'll break down how to craft high-converting ads using a proven Paid Social Ad Formula – Hook → Problem → Solution → CTA – as our guide. By understanding visual hierarchy, the use of headers, the primacy of message over polish, and the balance between quantity and quality, you'll learn how to tell a compelling story that captures attention and drives action. Let's dive in!

Visual Hierarchy Breakdown

Before we dive into a real-world example, let's look at a simplified version of how people typically consume ad content on social media:

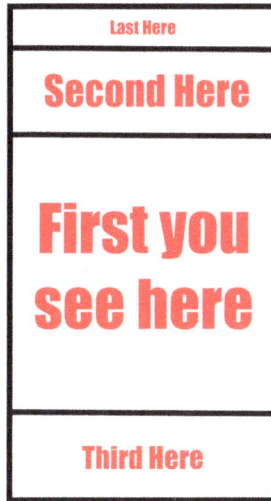

Visual Hierarchy

In most ad formats (especially static images or short videos), the viewer's attention follows a predictable visual path:

1. **First you see here** – This is the **main visual**, right at the center. It's the attention-grabber. Think of it as your *hook or Opener* — the first thing that either stops the scroll or doesn't.
2. **Second Here** – Once the viewer is intrigued, their eyes travel to the **next most visually dominant element**, often a header or secondary line of text. This can serve as the **problem** or context builder — the "why should I care?"
3. **Third Here** – Then the eye moves toward the **action prompt** — usually a button, offer, or product shot. This is the **solution/CTA** — what the ad is offering and what you should do.
4. **Last Here** – Finally, the eye may drift toward **supporting copy** — something smaller like "Free Shipping," social proof, or a bonus offer. It reinforces the message but never distracts from the core.

This hierarchy — **Visual → Message → Action → Details** — isn't just good design, it's cognitive science. It aligns with how our brains are wired to process information fast. You don't want viewers to work for your message. You want to guide them through it — instinctively and efficiently.

Now, let's look at how this plays out in a real example:

Ad retrieved from Meta Ad Library. Used for educational commentary.

The Dr. Squatch static ad example: it shows how a bold visual (the man and woman on the bed) and a provocative header text ("Best friend's gf won't stop smelling me ") immediately grab attention. This ad uses visual hierarchy to guide the viewer from the eye-catching opener to the call-to-action.

When a user encounters an ad – whether it's a **short video or a static image** – their eyes don't absorb everything at once. Instead, they follow a **visual path** through the content. A well-designed ad guides the viewer's gaze in a specific sequence, ensuring the key message hits home. Let's use the Dr. Squatch ad (pictured above) as a real-world example of how users' eyes typically travel through an ad's elements in order:

1. **Opener Image:** The *attention-grabbing visual* is seen first. In our example, it's the image of a man and woman on the bed. This unusual or intriguing scene is the **Hook** – it stops the scroll. (Fun fact: the human brain processes images *around 60,000 times faster* than text, so a striking image instantly draws attention.)

2. **Header Text:** Next, the viewer's eye catches the big, bold headline text overlay. Here that's the provocative **"Best friend's gf won't stop smelling me "** line. This text is deliberately large and high-contrast, so it pops out immediately. People naturally read *big text first*, even out of order, so a punchy headline can captivate the user into reading more. It succinctly presents a **Problem** (or at least teases a scenario) in a way that arouses curiosity.

3. **CTA Button:** Then attention moves to the call-to-action (CTA) – often a brightly colored **"Shop Now"** button or similar. Good ad design makes the CTA button visually distinctive (for instance, using a contrasting color) so that it *pops* on the screen. In a static ad, the CTA might be noticed third as the viewer looks for what to do next (in a video, the CTA button may appear throughout or at the end, but the viewer's intent to act is built by this point). This is the **Solution** and **CTA** part of the formula – the ad is prompting the viewer to take action (click, buy, learn more) as the solution to the teased problem.

4. **Supporting Text:** Finally, any supporting copy – for example, a smaller tagline like *"Only 321,169 men know this secret"* at the bottom – gets read. This text is usually last in the visual hierarchy because it's smaller and less contrasty. Its role is to reinforce intrigue or provide a bit of context after the main message. By the time someone reads this, they've likely been hooked by the header and are considering the solution, so this line nudges them further ("What secret? I want to know!"). It's an extra bit

of **substance** that can seal the deal, but it shouldn't distract from the main headline or CTA.

This visual hierarchy – **opener → headline → CTA → details** – works hand-in-hand with the Hook/Problem/Solution/CTA formula. The opener image and header together serve as the **Hook** (and hint at a Problem), the product offer implied by the ad is the **Solution**, and the button is the **CTA**. The magic is in presenting these elements in the right order. By leading the viewer's eyes through a story, you minimize the effort needed to understand the ad's message. In practice, a strong visual followed by a clear headline makes the viewer pause and *pay attention*, while the CTA and support text tell them what to do next and why.

Now, whether it's a five-second video or a single image, the same principles apply. **Short video ads** often mirror this structure: in the first 1–3 seconds, they flash an arresting scene or thumbnail (the opener) *and* overlay a bold text hook (the header) – for example, the video might open with the caption "I can't believe this trick worked " while showing a surprising clip. Immediately, the viewer is drawn in. As the video continues (seconds 3–5), it might introduce the **Problem** ("Struggling with X?") and then the **Solution** (the product or idea) shortly after, ultimately ending with a **CTA** ("Visit the link now" or a visible button on-screen). The key is that **the opener and headline hit together, right up front** – in social media, if you don't intrigue someone almost instantly, you lose them. Thus, in both static and video formats, the **first thing** a user sees should be something compelling enough to stop them in their tracks. Everything that follows should naturally flow from that hook, maintaining interest and delivering the payoff (solution and action).

Why does this structure work?

It aligns with how our brains and eyes process information. We latch onto visuals and bold text extremely quickly (our eyes follow an F-pattern or Z-pattern scan on screens). A flashy image or big headline at the top ensures the ad's most important point isn't missed. By the time the viewer's gaze moves down to the finer print, they're already invested. In essence, you are **telling a story in split-seconds**: first *grab attention*, then *deliver the key message*, then *invite action*, and lastly *provide any extra persuasion.* Mastering this visual storytelling is the foundation of high-converting ad creatives.

Should You Use a Header?

By now, you might be wondering, *"Do I always need a big headline text on my ad?"* In most cases, **yes – a concise, bold header can significantly boost an ad's performance.** If you scroll through a library of top-performing ads, you'll notice a pattern: many of them include a prominent text headline overlay. There's a good reason for that. A header instantly communicates the main idea or hook of your ad at a glance. It can either clarify what the viewer is looking at or intrigue them to find out more. Remember, people are scrolling fast – a punchy one-liner in **large font** can grab them before they swipe away. As one marketing guide puts it, having a strong headline at the top is a common way to captivate users and draw them into your ad's information. In short, the header often *is* the hook in text form, working with the imagery to stop the scroll.

That said, there are a few situations where headers shine and a few where they might not. Let's break it down:

- **When a Header Helps:** Use a header if you need to **deliver a message quickly** or create intrigue. For example, in our Dr. Squatch ad, the bizarre statement *"Best friend's gf won't stop smelling me "* on the image is doing a lot of heavy lifting – without it, the viewer sees two people on a bed and might be confused. The text instantly adds context (it hints at a *scenario* and piques curiosity – why is the friend's girlfriend sniffing him?). Headers are great for highlighting a **value proposition or problem** that the image alone can't show. If you're advertising a solution, a short headline can *frame the problem* for the viewer (e.g. "Tired of acne breakouts?" over a before-and-after picture). In user-generated content (UGC) style videos or TikToks, you'll often see creators literally put the hook text as a caption on the video within the first seconds – it's the same idea. A clear, readable header ensures the viewer immediately knows what's up or gets intrigued to keep watching. It's especially useful when your **visuals are busy or not self-explanatory**; a few words can make the difference between "Huh, what's this?" and "Oh, this is about solving *that* problem." Moreover, a header can create an **emotional trigger** or call out the target audience (e.g. "Calling all new moms – you need to see this"). This kind of direct approach often increases engagement because the right people notice it's meant for them.

- **When a Header Might Hurt (or Distract):** On the flip side, not every ad needs a big text overlay. If your visual is extremely clear or powerful on its own, adding text could be redundant or even distracting. There is such a thing as *too much* text on an image. Studies by Facebook have shown that ads overloaded with text tend to perform worse than those with a cleaner look. If your header text covers more than, say, 20% of the image, it might start to feel like a blatant ad and turn people off (back in the day, Facebook even had a 20% text rule for this reason). The goal is to **entice**, not overwhelm. So if you do use a header, keep it short – often one line or a few words is best. Another case where a header could be counterproductive is in highly **native or aesthetic content**. For instance, on Instagram, an ad that looks like a beautiful, organic post (with maybe just a small text in the image or none at all) can sometimes perform well because it doesn't immediately scream "advertisement." A giant headline in such a scenario could break the illusion. Similarly, in formats like Instagram Stories or Reels, you might choose to use the platform's native text stickers or subtitles rather than a huge banner text, to keep the style consistent with user content. And of course, if your ad is just a **simple demonstration** (imagine a 5-second video of someone using a product with a voiceover), plastering text might not be necessary if the voice or visuals already convey the hook. In short, use a header when it adds clarity or intrigue, but avoid it when it clutters the ad or states the obvious.

The main takeaway: **most of the time a headline will help more than it hurts**, especially in paid social ads. It's an opportunity to deliver your message in that critical first second. Just make it count – **clear, bold, and succinct.** Pair it well with your visuals (consider placement and font so it's readable instantly). And remember, header text and opener imagery usually appear together right away – particularly in short-form video, you typically see the person or product *and* a caption on screen in the first 3 seconds, working in tandem to hook the viewer. When done right, a header + visual combo can communicate the Hook/Problem portion of your formula almost telepathically fast. When done poorly, it can look like a wall of text. So craft those few words carefully and test different variations. It often makes the difference between an ad that *converts* and one that falls flat.

Message vs. Polish

It's easy to assume that the prettiest ads – the ones with cinematic footage, perfect lighting, and fancy effects – always perform best. But here's a comforting secret for aspiring editors on a budget: **a clear, compelling message will beat beautiful, flashy editing more often than not.** Substance trumps style in the land of high-converting ads. This is why we see an explosion of *raw*, "homemade" looking content in social advertising. Audiences have started to tune out ads that feel too polished or salesy; instead, they respond to content that feels **authentic and relatable**. In fact, authenticity has become so important that many brands prefer **UGC (User-**

Generated Content) style ads precisely because they look *less* professional – more like something your friend would post than a corporate ad. And people trust that. According to recent marketing insights, a whopping **92% of consumers trust user-generated content more than traditional, polished advertisements**. That trust translates to better performance: UGC-style ads often drive higher engagement and conversion rates than slick studio ads. Why? Because when an ad looks like a genuine user's post or an honest testimonial, viewers let their guard down. They think, consciously or not, "this seems real," and they pay attention to the **message** being conveyed.

For example, imagine two ads for the same skincare product: one is a glossy video with actors, studio lighting, and a scripted feel; the other is a selfie-style clip of a real person (or someone who feels real) talking excitedly about how the product cleared their skin. The first might be *beautiful*, but the second likely feels *believable*. The relatability and clear message ("this product solved my problem") outshine production value. In the social media era, **authentic storytelling** beats Hollywood cinematography for driving conversions. As one agency observing this trend noted, unlike high-end commercials, UGC ads leverage authenticity to forge a deeper connection with audiences – consumers increasingly value *honesty over perfection*. We've all seen TikTok videos or Instagram Stories shot on a phone that go insanely viral or make us want to buy something, precisely because they feel like content from a peer, not a polished ad. The takeaway: **don't be afraid if your ad "looks" simple** – worry more about whether it's communicating something that will resonate or solve a viewer's problem.

This isn't to say there's no room for professional production. Polish has its place, especially for branding and certain platforms. But even when you have high production values, you want to channel that authentic vibe. Many brands now use *UGC creators* or micro-influencers as "actors" – basically people who can speak to camera in a genuine way – instead of traditional actors with stiff scripts. The difference is noticeable: creators come off as "everyday people" sharing a tip, whereas a traditional actor might inadvertently give off a commercial vibe. (One comparison found that 48% of consumers prefer to follow influencers who **"look like everyday people,"** underlining how important relatability is.) The best scenario is when you combine a clear, strong message **with** an appealing presentation – but if you have to choose, choose message. A viewer might forgive a less-than-perfect edit or an iPhone-quality video if what's being said or shown interests them. They won't forgive a boring or irrelevant message just because you added cool graphics.

So, as a creative, focus on **clarity and authenticity** in your storytelling. Ask yourself: *What problem am I highlighting, and is the solution (product/service) obvious?* Does the ad feel like a *human* story rather than a corporate pitch? These factors matter more than having the slickest transitions. It's telling that many big advertisers have deliberately dialed down the "ad-iness" of their content. They incorporate ums and ahs, they use real customer testimonials, they embrace a bit of imperfection – because it feels **real**. And realness sells. Audiences are savvy; they've seen thousands of ads, and anything too polished can trigger that instinct to skip or doubt. But show them something that looks like it could've been made by a friend or a fellow user, and they're more inclined to trust it. In practical terms, this means **UGC is king** right now for performance marketing. Even if you're using actors or models, have them act **natural**. Even if you're editing in Premiere Pro, maybe make it *look* a bit like it was cut on TikTok. It also means **don't over-edit** – you might not need that crazy motion graphic or that perfect color grade; a slightly rough cut that gets to the point could do better.

Lastly, a quick note on **UGC talent vs. real users**: There's a spectrum. On one end, you have real customers creating content out of genuine love (gold if you can get it!). On the other, you have hired actors performing a script. A nice middle ground that many use are content creators (sometimes called UGC creators) – they are paid to produce content, but they do it in their own personable style. These folks usually appear more authentic than a traditional actor reading lines, because they speak in a more genuine tone and often film themselves in everyday settings. Consumers can tell when someone is *truly* into the product versus just paid to smile – in one study, over 20% of consumers said they'd unfollow an influencer who doesn't feel authentic or who promotes products they don't actually care about. The lesson for ad creatives: if you're scripting or guiding talent, **keep it real**. Encourage a natural tone, even if that means some slight messiness. The credibility you gain is worth its weight in conversions. In summary, **prioritize message over flash.** A clear value prop, a relatable story, and authenticity in style will take you much further than a video that's technically perfect but soulless. Substance is what convinces people to click "Buy Now." The style is just the vehicle – make sure that vehicle doesn't overshadow the message it's carrying.

Quality vs. Quantity & Quick vs. Clean

Now we come to a practical dilemma every creative faces: should you **crank out content quickly to test as much as possible**, or take your time to **polish each ad to perfection**? In other words, how do you balance *quantity* versus *quality* in producing ads? The honest answer: it depends on the situation. Both approaches are important in different moments, and a truly skilled video editor knows *when to move fast* and *when to slow down*. Let's talk about both sides and how to find a balance – while also keeping *yourself* sane and avoiding burnout (yes, we care about your health, too!).

When to prioritize speed and volume:

In the world of paid social, **iteration is key**. Often, you won't know exactly what will hit with your audience until you test it. This is why many advertisers favor a "quantity" approach – produce a bunch of ad variations, each with different hooks or angles, and see which one sticks. If you're working with a client or brand that provides a clear brief and lots of raw footage or ideas, it's usually a green light to execute quickly. For example, if the client says "we want four variations of this ad, each with a different hook image and a slightly different opening line," you don't need to spend weeks on each – you should **ship them fast** and get data. In scenarios where the strategy is set and it's more about tweaking or repurposing content, quantity can trump painstaking quality. Another case for speed is when **timeliness matters**. Say there's a trending meme or a seasonal event (Black Friday, New Year, etc.) – you might need to ride that wave *now*. In those cases, getting a decent ad out there quickly (while people still care about the trend) is better than spending so long perfecting it that you miss the moment. Additionally, consider the platform dynamics: social media feeds are ephemeral. An ad's lifespan might be just a few weeks before it fatigues. So pumping out new creatives regularly is necessary for consistent results. If you know you'll need 50 new ad creatives this quarter, you can't treat each one like a months-long film project. **Efficiency** is your friend. Use templates, reuse successful structures, and don't overthink minor details. Often, "good enough" in 2 days beats "maybe 5% better but took 2 weeks." In fact, focusing too much on pixel-perfect quality can lead to diminishing returns – time you could've spent making a *different* ad to test. Remember the saying: *"Done is better than perfect."* In many fast-paced marketing teams, that's basically a mantra. Quantity gives you more shots on goal, and more chances to find a high-converting winner. Especially when you have robust ad spend to quickly gather data, the quicker you can produce variations, the faster you learn what messaging or style resonates.

When to prioritize quality and thoroughness:

On the other hand, there are times when you **should slow down** and really put polish and thought into your ad. Not every situation allows for throwing spaghetti at the wall. Imagine you're working with a smaller brand with a limited budget – they can only afford to run one or two creatives this month. In that case, you want those few ads to be as strong as possible; you may not have the luxury of volume testing. This is where you invest extra time to develop a novel concept, write a sharp script, carefully edit, and refine the cut. Another scenario: you're introducing a **completely new concept or product** where the messaging isn't fully fleshed out yet. You might need to brainstorm and script multiple ideas internally, but when it comes to production, you focus on one or two best concepts to execute with high quality. Essentially, if you're breaking new ground, *quality* of idea and execution matters more than sheer quantity. Additionally, certain campaigns (like a big product launch or a brand's signature ad) warrant more TLC. You wouldn't rush out the Super Bowl commercial in a day, right? For these, taking the time to ensure everything is just right – the story flow, the visuals, the sound mix – can pay off in a big way, because these ads carry more weight. There's also the matter of **craftsmanship**. As an editor, you'll sometimes see opportunities that require a bit more work but can significantly boost performance – for instance, digging through a ton of raw footage to find *the* most thumb-stopping opening shot. This takes time. Or maybe manually adding captions to a talking video to ensure they're perfectly synced and styled (auto-captions might be fast, but a custom touch could make it more engaging). When you have hints that an ad concept is promising, *that's* when polishing it up can turn a good ad into a great ad. One more angle: **quality can mean ensuring your message is crystal clear.** Rushing might cause you to put out an ad that is confusing or not as compelling as it could be, which is a wasted opportunity. Taking an extra day to refine the copy or structure could. dramatically improve its effectiveness – which matters if you're only launching a few ads.

In practice, you'll often do a bit of both: pump out a batch of ideas, then identify a couple that show potential and refine those further. It's a cycle of **test → learn → optimize**. For instance, you might quickly produce five video ads with slight differences in hooks. After a week of data, two of them clearly perform best. You could then take those two and make *improved* versions (better editing, maybe combine elements from the losers that had merits, etc.), investing more time now that you have evidence they're worth it. This way you get the best of both worlds: the exploratory power of quantity and the enhanced results of quality.

While juggling this, it's super important to **set personal boundaries and avoid burnout.** The advertising industry moves at lightning speed, and creatives are often under pressure to do more, more, more – *yesterday.* But churning out content 24/7 at the cost of your mental or physical health is a recipe for disaster (for both you and the ads). If you're constantly rushing everything, not only will your work likely suffer, but you *will* burn out. And you're not alone – a recent survey revealed that about **70% of professionals in media, marketing, and creative fields have experienced burnout**. That's a sobering statistic, and it highlights how common and serious this issue is. To be a sustainable creative force, you have to balance hustle with self-care. What does that mean in context? It means communicating realistic timelines to your team or clients. If you know that producing 10 high-quality video ads in one week is humanly impossible, say so up front and help set priorities. It means establishing that sometimes you need a day to *just think* or brainstorm – which is part of the work, even if it doesn't result in an immediate deliverable. It also means using tools and workflows that speed up the boring stuff (like using templates, presets, or even AI tools for rough cuts) so you can focus your energy on the creative decisions rather than grinding 16 hours in an editor for a minor tweak. Crucially, don't be afraid to **push back or ask for help** if expectations are unrealistic. Setting boundaries might sound scary, especially early in your career, but it will actually improve your output in the long run. A well-rested, focused editor will produce far better work (and more of it) over time than a burnt-out one running on fumes.

To wrap up, think of *quality vs. quantity* as a dial you tune based on context. If you're in a rapid testing environment with ample budget – lean towards **quantity** (with a baseline standard of quality, of course). If you're in a high-stakes or constrained scenario – lean into **quality** for each piece. And often, the strategy is to iterate quickly, then refine: speed when conceptualizing and testing, care when polishing the winners. Through it all, keep the big picture in mind: a high-converting ad isn't born from either spamming dozens of mediocre videos *or* spending months on one masterpiece; it often comes from a smart process that uses both approaches. **Be fast when you can, and be**

thorough when it counts. And above all, **take care of your creative spark** – it's the engine of all your high-converting ideas. Protect it by finding that balance in your workload. Your future self (and your clients and audiences) will thank you.

Keep it up! By understanding structure, style, and substance – and how to balance all three – you're well on your way to creating ads that not only look good and feel genuine, but also drive real results. Every ad you make is an opportunity to learn. So experiment boldly with hooks and headlines, focus on messages that matter, and refine your craft one edit at a time. High-converting ads are as much an art as a science; with the principles from this chapter in your toolkit, you have a solid foundation. Now go forth and create something awesome – and don't forget to have a little fun with it, too!

8

Performance Craft – The Techniques Behind Converting Cuts

Once you've built a solid structure and nailed your messaging, it's time to sharpen your **editing craft**. This chapter is all about using creative techniques—**the actual cuts, timing, and visual tricks**—that increase engagement, improve retention, and boost conversion.

These aren't about making your edit *look cool* — they're about making people *keep watching.*

Your goal isn't art. It's attention.

Why Editing Techniques Matter in Paid Social

Social media feeds are brutal. People scroll with one thumb while half-watching a show, half-listening to a podcast, half-having a conversation. If you want them to stick with your video, **you need to control pacing, re-engage attention, and reduce *drop-off*.** Editing techniques are your toolkit to do just that.

Let's break down the most effective ones used in performance ads:

Jump Cuts – Keep It Moving

Jump cuts are the fastest way to eliminate dead space. You use them to:

- Remove silences and filler words ("uhh," "so...")
- Speed up a talking head to keep energy up
- Emphasize changes in expression or movement

Why it works: Jump cuts add rhythm. They create **micro-movement** that keeps the visual pace snappy. In platforms like TikTok, Meta, and YouTube Shorts, stillness kills. Every cut is a chance to reset attention. Even when editing testimonials or UGC, jump cuts are essential to **compress the story** into a tight, watchable edit.

Zooms & Punch-ins – Add Focus & Drama

Digital zoom-ins (also called punch-ins) are used to highlight a facial expression or a keyword. They're great for:

- Emphasizing key phrases ("...and it *actually* worked!")
- Guiding the viewer's eye
- Breaking the monotony of a static frame

Why it works: When the face gets closer, the viewer feels more emotionally connected. A zoom suggests **importance**. Use it to bring a viewer into the moment without needing a new shot. It's especially useful for talking-head videos where you only have one angle.

Captions – Your Safety Net

Every video ad should have **on-screen text** or captions. Whether it's auto-captioned or stylized, make sure the message can be read even with the sound off.

- Use dynamic captions that match the speaker's cadence
- Highlight important words with bold colors, movement, or scaling
- Avoid tiny text or low contrast

Why it works: Most people **watch with no sound**, especially on Meta and Instagram. Captions ensure your message is absorbed. Also, stylized captions add motion—another way to re-capture attention.

Sound Design – Don't Just Add Music, Design It

Sound can increase the **emotional impact** of an ad.

- Use music that **matches the tone** of the product and the edit pace
- Add **subtle sound effects**: swooshes, clicks, whooshes, pops
- Sync **beat drops** or lyric changes with big visual moments (e.g. product reveal, CTA)

Why it works: Great sound makes visuals feel more polished and intentional. It **keeps the energy up**, builds anticipation, and can even create recall ("that ad with the beat drop when the product appeared").

Boomerangs – Quick Loops to Reinforce Visuals

Boomerangs (very short clips that loop back and forth) can be surprisingly effective when used intentionally.

- Loop a dramatic visual to **re-emphasize a point**
- Use for **before/after gestures**, product reactions, or eye-catching motions

Why it works: The loop gives the brain **two chances to register** what just happened. It adds novelty and a slight "what did I just see?" moment that creates curiosity. You don't need to overdo it—but as a punch in an opener or CTA, it can work.

Speed Ramps – Control the Flow

Speed ramping = slowing down or speeding up footage at specific points.

- Slow motion on product use or reactions
- Speed up setup or boring parts
- Add acceleration before a transition or beat drop

Why it works: Controlling tempo lets you **highlight key moments** and avoid viewer boredom. It gives your video a flow that mirrors music and storytelling beats. It's especially useful for visual-first platforms like TikTok and Instagram Reels.

When to Use These Techniques

Don't use all of them at once. Let the **content and message** guide you.

For UGC and testimonials: Jump cuts, captions, punch-ins are your go-to. Keep it tight, clear, and dynamic.

For product demos: Sound design, speed ramps, boomerangs help dramatize moments and show benefits clearly.

For story-driven edits: Structure and pacing matter most. Use techniques to keep people watching **past the opener**, through the **onramp**, and into the **sales sequence**.

What You're Really Doing: Retention Editing

Let's be real—most editing in paid social is about **keeping people watching**. Every few seconds, you're asking:

"Am I losing them?"

These techniques are like little tools in your belt to **pull attention back in**.

You're not just cutting a video.

You're building **momentum**.

You're shaping **emotion**.

You're playing with **curiosity** and **pattern**.

That's what makes an editor more than just a technician. It makes you a creative performance partner.

9

Platform-Specific Secrets

Not all platforms are built the same—and neither are the ads that win on them.

If you try to edit a TikTok ad like a YouTube video, it'll flop. If you post a Meta ad with a YouTube-style intro, people scroll right past. Each platform has its **own tone, viewer behavior, and algorithm preferences**, and your job is to adapt your editing to meet them.

This chapter breaks down the creative rules (and smart exceptions) for the three biggest platforms: **Meta (Facebook/Instagram), TikTok, and YouTube Shorts**.

Meta (Facebook & Instagram)

Meta is the **OG of paid social**—and still one of the most effective platforms for direct response and ROI-driven campaigns. Here's what works:

CTA-Driven Edits

Everything on Meta should push viewers toward action: clicking, swiping, or buying. That means:

- Strong CTAs visually and verbally
- Direct benefit-driven headers
- Shorter intros (get to the value fast)

Example:
  A skincare ad might start with:

> *"Still wasting money on drugstore serums?"*
>   ...followed by a hooky visual and a *"Shop Now"* button within 3–5 seconds.

Performance-First, Not Trend-First

Unlike TikTok, Meta doesn't reward trend-jacking as much. Instead:

- Focus on **UGC remixes**: real people, real problems, real results
- Keep **copy clean and bold**: big fonts, white space, emojis
- Use **square (1:1)** or **4:5** for best feed performance and **Vertical (9:16)** for reels and stories
- Use **native captions** and auto-play sound to your advantage

What to Watch For:

- **Thumbstop Rate**: Did they stop scrolling?
- **CTR**: Are they clicking?
- **Hook Rate / Hold Rate**: Are they watching past 3s / 6s?

Meta gives you strong performance data—**test more, iterate faster**.

TikTok

TikTok is its own language. And if you speak it fluently, you can make ads feel like native content instead of... ads.

Native Cuts Win

TikTok is raw, fast, and emotional. That means:

- Cut to the hook **immediately** (no intros)
- Use **sound bites** or trending audio (but avoid relying on trends for paid ads unless part of strategy)
- Embrace **imperfect**: lighting, camera shake, real talk wins
- Use **text overlays** the way creators do—caption-style, not fancy motion graphics

Example:

> "I thought this was gonna be another BS fat-burner... but then I tried it for 30 days."

That feels native. It feels like a post, not a pitch.

Shorter Is Often Better

15–25 seconds tends to win. But strong hooks can stretch to 45–60s if the story is engaging enough.

Storytelling Structure That Works:

- **Relatable hook** ("I didn't believe it would work...")
- **Product intro/demo**
- **Problem + transformation**
- **Social proof or humor**
- **Call-to-action or emotional close**

What to Watch For:

- **Hook Rate**: Are people watching the first 3s?
- **Hold Rate**: Do they stay till 6s?
- **Engagement Rate**: Likes, comments, shares—your content shouldn't feel like a brand ad

If the content **feels like a friend posted it**, you're on the right track.

YouTube Shorts

YouTube Shorts is newer in the ad space—but growing fast, especially for **top-of-funnel brand awareness** and **educational content**. Here's how to make your creative fit in:

Treat It Like Mini-YouTube, Not TikTok

- Viewers are used to **narrative** and **explanation**
- Go slower with **context and clarity**
- More **voiceovers**, less jump-cut chaos

You can still hook fast, but you can also breathe a little more.

Example:

> "This is the one mistake 90% of people make when brushing their teeth..."
>> Let that line breathe. Then explain. Then show. Then CTA.

Design for Playback, Not Scroll

Unlike Meta or TikTok, users may **rewatch** content. So:

- Keep the **pacing clear** (don't rely on flashy transitions)
- Re-introduce product name or brand 2–3 times
- Think **educational first, emotional second** (this works well for SaaS, coaching, health)

What to Watch For:

- **Midpoint Retention**: Did they make it past 50%?
- **View Duration**: YouTube cares about how long they stay
- **Click-Through Rate** if paired with CTA buttons (for Shorts ads)

YouTube Shorts favors ads that **teach, inspire, or dramatize value**.

Platform Recap

Platform Recap			
Platform	Focus	Style	Best For
1 Meta (IG/FB)	CTA-first, scroll-stopping	Clean, UGC-style, bold text	Direct response & retargeting
2 TikTok	Native & emotional	Raw, fast, creator-first	UGC, testimonials, viral hooks
3 YouTube Shorts	Narrative & retention	Slower pacing, informative edits	TOF education, SaaS, coaches

Your Superpower: Platform Fluency

Great performance editors don't just "make cool videos." They **speak the language of each platform**, and they translate creative ideas into native formats that blend in—*but convert.*

You'll never treat Meta like TikTok again. Or YouTube like Meta.

And that's what makes you irreplaceable.

10

Advanced UGC Editing – The Ultimate Guide to a High-Converting Ad Archetype

User-Generated Content (UGC) style videos have emerged as the **most important archetype in paid social ads** today. In this chapter, we'll build on the foundations from Chapter 2 and dive into advanced techniques for editing UGC ads. By now, you know that authentic, relatable content can hook a viewer faster than any high-gloss commercial. Here, we'll provide a step-by-step guide and instructional overview on executing UGC-style ads at an advanced level – from structuring your content for maximum retention to mining unexpected footage for hidden gold.

UGC ads thrive on authenticity and storytelling. They often **outperform traditional ads** by significant margins – for example, UGC-based ads can achieve **4× higher click-through rates** and **50% lower cost-per-click** than average ads. And it's not just the numbers; marketers have taken note as well. In fact, **93% of marketers who leverage UGC say it performs better than traditional branded content**. Why is that? Viewers trust content that feels real. Studies show **84% of consumers trust peer recommendations above all other forms of advertising**, so an ad that feels like a friend's honest recommendation can be incredibly powerful.

In Chapter 2, we introduced the basic UGC structure – hooking the audience with genuine stories and delivering a solution through authenticity. Now, in Chapter 10, we'll expand on those concepts with a detailed look at the **four major UGC content sections** and how to expertly sequence them for retention and engagement. These four sections are:

1. **Personal Context** – Presenting relatable problems, desires, or personal history to create an emotional hook.
2. **Service Info** – Introducing the brand or product as the solution, including how it works, special offers, or a call-to-action.
3. **Product Interaction** – Demonstrating the product or service in use (unboxing, demo, sensory details, key features) to build credibility.
4. **Results** – Showcasing outcomes (personal results, reviews, "after" scenes, or favorites) to provide proof and payoff.

We'll explore each section in detail with clear examples and best practices. You'll learn how to **structure these sequences** into a cohesive narrative that keeps viewers watching and drives them to take action. Consider this chapter your advanced playbook for UGC editing – by the end, you should be able to turn raw footage into polished, high-converting social ads. Let's get started!

1. Personal Context – Hooking with a Relatable Story

Before we dive in, let's get one thing straight: **the Hook is *not necessarily* the same as the Personal Context.**

The **Hook** refers to the very first few seconds of your ad – typically the **Opener** (visual/audio) and **Header** (on-screen text or headline) – designed to stop the scroll and earn attention. The **Personal Context** usually follows the hook and sets up the deeper story. Together, the Hook + Personal Context form what we'll refer to as the **Onramp** – the crucial opening stretch that brings viewers into the ad experience.

That said, in **most high-performing UGC ads**, the Hook *is* often connected to or inspired by the Personal Context – but it doesn't have to be. You can open with a bold statement, a surprising visual, a stat, a meme, or even the *result* – then backtrack into the story. The key is that once the Hook grabs attention, the Personal Context quickly makes the viewer care enough to keep watching.

What is Personal Context?

The **Personal Context** section highlights a problem, desire, or backstory that your target audience can relate to. It's answering the viewer's unspoken question:

> **"Why should I care?"**

A strong personal context creates empathy and draws the viewer in because they see themselves in the story being told. In UGC ads, it often comes through as someone speaking candidly about their real-life experience.

In Practice: Hook vs. Personal Context

Let's break it down:

Hook (Opener + Header):

- First 1–3 seconds.
- The attention-grabber.

Could be:

- A dramatic result ("Lost 5 lbs in 2 weeks?!")
- A visual gag

- A meme format
- A high-contrast before/after
- A striking statistic or quote

Personal Context:

Comes right after. This is where the storyteller starts to share their situation:

- Their struggles
- Their lifestyle
- The "before" picture
- A relatable emotional state

Together, they form the **Onramp** – the most important 5–7 seconds of your ad.

Example:

Hook (Opener + Header):

> On-screen text: *"I was THIS close to deleting every fitness app I owned."*
> Clip: Woman frustrated, deleting apps on her phone.

Then... Personal Context:

> "After my second child, I barely recognized myself. I was exhausted, out of shape, and honestly a bit depressed about it."

Notice how the Hook sets up curiosity and relatability immediately. The Personal Context then builds the emotional connection, setting up the rest of the story.

Core Elements of a Strong Personal Context:

The Problem or Pain Point:

- "I couldn't lose the baby weight no matter what I tried."
- "My old vacuum was so heavy it hurt my back whenever I cleaned."

Emotional Resonance: Show frustration, disappointment, hope – with both words and visuals.

Brevity and Clarity: Get to the core message fast. Cut the fluff.

> **Pro Tip:** Your Hook should stop the scroll, but your Personal Context is what earns the right to keep watching. Make sure the transition from one to the other feels seamless, and that the combined Onramp clearly signals what the viewer is in for – without overexplaining.

Authenticity wins here. Keep it raw, conversational, and human. Minor imperfections? They actually help.

2. Service Info – Introducing the Solution (Brand & Offer)

After establishing the viewer's problem or desire through personal context, the next logical step is to introduce the **solution** – and that's where the **Service Info** section comes in. In this part of the UGC ad, you present the product or service (and the brand behind it) as the answer to the problem, and you provide key information that the viewer needs to know. This includes **who** the brand is, **what** the product/service does, **how** it works, and often an incentive or **call-to-action** to nudge the viewer towards conversion.

Think of Service Info as **connecting the dots** for the viewer: "Here's what can solve the problem you just saw." It's important to transition smoothly – the viewer should feel a natural progression from the story ("I had this issue...") to the introduction of the product ("...so I tried [Brand/Product X]"). Often a simple connector line is used, such as *"Then I found **Product X**,"* or *"That's when I discovered **Brand Y**."* Immediately after, the ad should briefly **introduce the product/brand** in a positive light.

What to include in Service Info:

- **Brand/Product Name and Type:** Clearly state what it is. ("**FitLife App** – a personal trainer in your pocket," or "**SparkClean Vacuum** – a lightweight wireless vacuum.") This ensures viewers know what's being advertised.

- **Unique Value Proposition:** Why is this product special or tailored to the problem? This could be a feature or an approach. ("They even offer a free quiz to customize your workout plan," or "It's 50% lighter than standard vacuums, making cleaning effortless.")

- **How It Works:** Especially for services or apps, give a quick summary of the user journey. ("I downloaded the app, answered a few questions about my goals, and it gave me a 4-week fitness plan," or "It runs on a rechargeable battery – just charge, press a button, and start cleaning, no cords needed.") Keep it **simple and digestible**.

- **Offer or CTA (Call-to-Action):** If there's a special offer, mention it here or towards the end of this section. ("They're offering 1 month free for new users," or "I used their code for 20% off my first order.") Even if there's no special deal, you can encourage the viewer: *"You can check it out via the link below"* or *"I had to try it myself to believe it."* This plants the seed that the viewer can take action too.

This section should **remain concise** – you're not dumping the entire website copy here. In a quick ad, you might spend only 5-10 seconds on the Service Info elements. The goal is to inform enough to give context, **build credibility**, and then quickly move on to showing the product in action. You want the viewer to understand what the solution is, but you don't want to bog them down with too many details or a hard sell (which can feel jarring in a UGC-style ad).

Example transitions into Service Info: Continuing the fitness app scenario from above, here's how the ad might transition:

- After the personal context hook (the mom expressing exhaustion and being out of shape), she might say: *"Then I heard about this new fitness app called **FitLife** from a friend."*

- On screen, as she says the name, you might overlay the app's logo or a screenshot for instant brand recognition.

- She could continue: *"It promised me customized workouts I could do at home in just 15 minutes a day."* – This line introduces the key benefit (short, customized workouts at home) which directly addresses her problem of being busy and out of shape.

- She might add: *"I was skeptical, but they had a free 1-month trial, so what did I have to lose?"* – Here we weave in an offer (free trial) and her thought process, which encourages viewers to have the same mentality ("nothing to lose by trying").

In that brief span, the viewer learned **what** the product is (a fitness app), **what it does** (short custom workouts), **the brand name** (FitLife), and even got an **incentive** (free trial). It's all done through the narrative of the storyteller's experience, which keeps it feeling natural and not like an advertisement blasting facts.

> **Note:** It's crucial that the tone stays conversational. Even though you are delivering "service information," it shouldn't feel like a formal commercial or a scripted sales pitch. In UGC style, the information is often delivered as part of the person's story ("I found this...", "I liked that it...", "they even offered..."). This maintains authenticity and trust. If the ad simply cuts to a different voice or a corporate-style blurb about the product, it will break the illusion of a peer-to-peer recommendation. So, have the *same person in the ad convey the service info*, or present it in a visually seamless way (like text bubbles or screen recordings) that feels like it's still coming from a personal place.

By the end of the Service Info section, your viewer should clearly know **what solution is being offered** and feel like it's a credible answer to the initial problem. They should also have a sense of *excitement or curiosity* about seeing it in action – which leads perfectly into the next section: **showing the product or service in use.**

3. Product Interaction – Show, Don't Tell (Unboxing, Demo, and Features)

Now that the viewer knows *what* the solution is, it's time to **prove that it delivers**. The **Product Interaction** section is where you bring the product or service to life on screen. This is a make-or-break segment for engagement: viewers want to see the **tangible reality** of the product – how it looks, how it works, and how it fits into someone's life. In a UGC ad, this portion often includes footage like unboxing, demonstrations, and close-ups, all presented in a relatable, non-commercial style.

The mantra for this section is **"show, don't just tell."** Think of it as the *evidence* supporting the claims made in the Service Info. It's one thing for our storyteller to *say* the vacuum is lightweight or the shake tastes great; it's another to *visually show* those qualities in action. Here's how to nail the product interaction section:

- **Unboxing / First Impressions:** If applicable, show the moment the product arrives or is opened. This can be quick, satisfying shots of opening a package, peeling off protective wraps, or the first look at the item. Unboxing scenes create anticipation and make the audience feel like they're right there sharing the experience. For example, *"A few days later, a box showed up on my doorstep..."* could be narrated over a clip of the package being opened to reveal the product neatly packed inside. If it's a service or app, "unboxing" might be showing the first time using the app or a welcome email, etc. The idea is to capture that initial excitement.

- **Demonstration / Usage:** Show the product being used in a realistic scenario. If it's a physical product, footage of the person actually using it is key (drinking the shake, working out with the app, vacuuming the floor, applying the skincare serum, etc.). If it's an app or service, consider

127

screen recordings or over-the-shoulder shots of the app interface, or footage of the person engaging with it (typing, interacting with on-screen elements). Pair these visuals with commentary that **highlights the experience**: *"It smelled amazing right out of the box"* (for a candle), *"The app's interface was super intuitive – it even talks you through each exercise"*, *"This vacuum glides so easily, I can do my whole living room in 5 minutes!"*. You're illustrating benefits through action.

· **Sensory and Details:** One advantage of video is appealing to senses beyond just describing features. If the product has a sensory element (taste, smell, feel, sound), try to convey that. The person might take a bite and go "Mmm!" to show taste, or rub the fabric to show softness, or mention "It's got this fresh citrus scent" while smiling. Use close-up shots for texture, color, or unique design elements. Even though viewers can't physically experience it, these cues help them imagine it.

· **Feature Callouts:** As the product is being shown, call out key features **in context**. This can be done through the narration ("...and it's waterproof, so I even used it in the shower without worry") and/or with on-screen text or graphics. A small caption pointing to the product saying "100% Waterproof" or "No sugar added" or "Made with recyclable materials" can reinforce the selling points. Keep these brief and *integrated into the video's flow* – they can appear as natural labels or highlights, not as formal bullet points. The idea is that the viewer's *eyes* can catch a feature highlight even if the audio is off or if they miss the spoken part. (In fact, remember that many viewers **watch videos on mute**, so visual reinforcement is crucial.)

Pro Tip: Edit this section with a focus on pacing and clarity. It can be tempting to include every single clip of the product, but you should choose the most impactful moments. Trim out any dead air or slow bits – **snappy pacing keeps viewers hooked**. At the same time, *don't rush so much that nothing is clear.* Each clip should last long enough for the viewer to understand what they're seeing. A good rule of thumb is to cut on action and keep things moving, but pause a half-beat on truly important visuals (e.g., a before/after comparison, or a crucial feature highlight).

Also, **use subtitles or captions for any spoken explanation of features** – not just because people watch on mute, but also to double down on communicating the value. If the demonstrator says, "It only took me 30 seconds to blend a smoothie," you might flash the text "Only 30 seconds to blend!" on screen. This ensures the message lands clearly.

Let's visualize the **Product Interaction** with our ongoing example scenarios:

- *Fitness App example:* Show the mom setting her phone on a table and doing a quick workout guided by the app. On-screen, we see a bit of the app interface (maybe the workout timer counting down from 15:00). We see her doing modified push-ups while the app's voice encourages her (if audible). She's smiling or at least looking determined. She might say in a voiceover, "The workouts are actually fun, and I can squeeze them in while the baby naps!" Meanwhile, text appears: "15-min Home Workout ✅ Fun & Quick" to underline the point. We might also show a close-up of her phone screen with the day's workout completed, or her tapping "Done" and looking happy.

- *Vacuum example:* Show a clip of the person easily lifting the vacuum with one hand (to demonstrate light weight), pushing it across the floor and it picking up visible debris. Perhaps include a before/after of a dirty carpet spot now clean. The narrator might say, "See that? It picked up *everything* in one pass." A text callout could say "Powerful one-pass cleaning" near the vacuum head. We might hear the vacuum running, then she turns it off and you can hear her say, "And it's so quiet my dog isn't even scared of it!" (If possible, a shot of a calm dog nearby for a touch of humor and relatability.)

This section is often the **longest segment** of the ad in terms of footage, but the narration or story should still be tight. Aim to keep it engaging by **varying the visuals**: alternate between the person using the product and product close-ups, switch angles, show reactions, etc. This variation acts as a pattern interrupt to re-capture attention if it started to wane.

At the same time, ensure **cohesion** – all these shots should feel like they belong to one continuous experience. Good editing can achieve this by using consistent lighting/color, keeping the person in the same outfit/environment, or adding a subtle background music track that ties the scene together. (Just make sure music *never overpowers* the speaking parts – it should complement, not distract.)

Finally, a word on **authenticity in the demo**: It's important that the Product Interaction doesn't suddenly look like a glossy commercial. Avoid overly staged scenes. It's okay if the camera is handheld or if the lighting is just normal room lighting (as long as the product can be seen clearly). The focus is on *genuine use*. If the person fumbles slightly while opening the box, that's fine – it can be endearing and real. You're showing a *real person interacting with a real product*, which builds trust. Remember, **don't over-edit** to the point of losing authenticity – too many flashy cuts or effects can actually turn viewers off and erode trust. Let the product and the person's honest reactions speak for themselves.

By the end of the Product Interaction section, your audience should be thinking, *"Okay, I see it working. It does what they said it does."* They've been on the journey from problem to solution to seeing the solution in action. Now they need one final push to solidify their desire to try the product – and that comes with the results and payoff.

4. Results – Delivering the Payoff (Testimonials, Outcomes & Favorites)

The final act of your UGC ad narrative is the **Results** section. This is where you show the viewer the *outcome* of using the product or service, wrapping up the story you started in the Personal Context. A well-crafted results segment provides **proof** that the solution truly works and leaves the viewer with a satisfying conclusion – ideally, feeling inspired or excited to achieve similar results themselves. In marketing terms, this is where you drive the conversion home by answering: "Did this solve the problem, and was it worth it?"

Key components of the Results section include:

- **Personal Outcome:** Continue the story of the individual featured. How did their life improve after using the product? Be specific and genuine. For instance, the fitness app user might say, *"Fast forward a month – I've lost 5 pounds and I have so much more energy!"* She could appear on screen looking happier, maybe showing a before-and-after photo or simply wearing an outfit that now fits better. The vacuum user might show their spotless living room and say, *"Now cleaning is a breeze, and my home is the cleanest it's ever been – without any back pain."* This closes the loop that was opened with the initial problem.

- **Testimonials or Social Proof:** In UGC-style ads, sometimes the person's own result is the testimonial. But you can bolster credibility by briefly showing *additional* proof points. This could be a quick collage of other user reviews or quotes (e.g., overlay text like "★★★★★ Thousands of happy customers" or a short clip of another person saying "It worked for me too!"). If the ad format allows, you might even do a rapid-fire sequence: "Here's what others are saying," followed by two or three one-liner testimonials or quick face-cam snippets: *"Best decision ever!"*, *"Saw results in just 2 weeks,"* etc. Keep it brief and positive. This signals that it's not just one anomaly – the product consistently delivers results.

· **Favorite Features or Summary:** Sometimes the person will cap off their story by reiterating what they love most. *"My favorite part? I actually enjoy the workouts now, and it fits into my day."* Or *"I love that I don't need to drag out a heavy vacuum anymore – this thing does the job and then tucks away easily."* This helps reinforce a key benefit as a lasting impression. It's basically the user saying: *"I'm a fan of this product, and here's why."*

· **Call to Action (CTA):** Every ad needs a call to action, and here is where it should be loud and clear. Because this is UGC style, the CTA often comes from the persona's voice or is shown in text overlay (or both). It might be as straightforward as: *"You've got to try **FitLife** for yourself"* or *"I highly recommend **SparkClean** if you hate lugging around old vacuums – check it out!"*. Additionally, if there's a link or code, mention it: *"Click the link to get started"*, *"Use my code **FITMOM** for a free trial"*, or *"Trust me, you'll wonder how you lived without it. Tap below to get yours."* The tone is encouraging and excited, not coercive. The viewer should feel like, *"Hey, this person had great results and is enthusiastically telling me to give it a shot"*, which comes off as friendly advice.

When editing the Results section, you want to achieve an **emotional high point**. This is the climax of the story: problem was overcome, life is better, and the viewer should feel a sense of resolution and optimism. Music (if you're using any) can swell a little here – nothing too dramatic, but a slight uplift to match the positive outcome can work nicely. Visually, this section can include bright, "after" shots: smiling faces, tidy spaces, healthy glow, etc., depending on the narrative. If you have a before-and-after contrast, show it (side by side or a quick cut from "then" to "now"). Seeing is believing.

Example – Tying it all together: Let's finalize our fitness app story in this Results segment:

- The mom now appears looking confident in workout clothes. She's holding her phone showing the app's results page or maybe a progress chart. She says, *"It's been 4 weeks and I feel like myself again. I've lost 5 pounds, gained a ton of energy, and I actually look forward to my 15-minute workouts!"* As she speaks, we might see a comparison of her Day 1 and Day 30 photos (if available) or even text on screen: "-5 lbs in 4 weeks ✔ More energy ✔".

- She continues with a smile, *"Honestly, **FitLife** made it so easy. I can't believe I waited this long to try it."* This serves as both praise and a subtle nudge to the viewer (don't wait, try it now).

- For extra social proof, the video could cut to a screenshot of a 5-star rating or a quick quote from another user review: *"Best fitness app ever – @jane_doe"*, just for one second, then back to our main person. (This is optional, but can reinforce that she's not the only one).

- Finally, the CTA: She looks directly into the camera and says, *"If you're a busy person who wants to get fit, you have to try* **FitLife***. Trust me, you won't regret it!"* The video might then jump into an **end card or outro CTA**, where the **brand logo** appears alongside the **promo message**: "**Try FitLife – Get your 1 month free →**" as the closing shot.

In those closing moments, the viewer has seen the full journey: a relatable struggle, the introduction of a solution, that solution being used, and the happy result. This narrative arc is satisfying to watch and also highly persuasive. We've essentially shown a mini "movie" of someone's transformation, and by doing so we addressed potential customer questions along the way (Is this for people like me? What does the product actually do? Will it work? What results can I expect?). By the end, a viewer who relates to that initial context is likely thinking, *"I want those results too. I'm going to click this ad."* That's the conversion moment.

> **Tip:** When showcasing results, always stay truthful. Don't depict unrealistic outcomes or guarantee something the product can't uniformly deliver. Viewers are savvy, and UGC is all about **trust**. It's far more powerful to have a modest but genuine result ("I feel more confident in my skin now" or "I save 2 hours every week on cleaning") than an unbelievable claim that triggers skepticism. Real people with real improvements make for convincing testimonials.

At this stage, you have guided your viewer through all four sections of UGC content: Personal Context → Service Info → Product Interaction → Results. Next, we'll discuss how to fine-tune the sequencing and some advanced editing tactics to ensure each part flows seamlessly into the next, keeping viewers engaged throughout the ad.

Crafting the Sequence for Maximum Retention and Engagement ·

Understanding each section of UGC content is vital, but **execution is all about how you stitch them together** into a compelling sequence. In a well-edited UGC ad, the transitions between Personal Context, Service Info, Product Interaction, and Results feel natural and keep the viewer hooked from start to finish. Let's explore some advanced tips for sequencing and engagement:

- **Follow the Story Arc (Problem → Solution → Proof):** The order of the four sections is designed to follow a classic narrative arc that humans respond to. Sticking to this flow (or a clever variation of it) ensures the ad makes sense and feels satisfying. The problem is introduced, a solution is offered, proof is shown, and the story concludes. Most high-converting UGC ads use this formula in one way or another because it mirrors how we *think* when making a decision (identify problem, consider solution, look for proof, then decide). That said, don't be afraid to **get creative** within this structure. For example, some editors start with a lightning-fast teaser of the result to spark curiosity ("Before we get into it, look at this after-shot... amazing, right? Now let me show you how I got there."). This can act as a hook before actually diving into the personal context. Use such techniques judiciously – they can boost retention by promising a payoff if the viewer keeps watching.

- **Smooth Transitions:** Each section should lead into the next like chapters in a tight short story. Avoid jarring cuts that feel like the topic is shifting abruptly. One technique is to use **bridging sentences or visuals**. For instance, at the end of the Personal Context, a phrase like *"...and that's when I decided to try [Product]."* cleanly opens the door for Service Info. Similarly, at the end of Service Info, something like *"A few days later, my order arrived..."* naturally sets up the Product Interaction/unboxing.

135

These little phrases act as signposts for the viewer, guiding them along. Visually, you can complement this by matching the cut with a related image (e.g., showing the product right as it's mentioned).

· **Pacing and Timing:** Modern social media users have *very short attention spans.* A critical advanced skill is learning **how long to spend on each section**. While there's no one-size rule, a common breakdown for, say, a 30-second ad might be: ~5 seconds Personal Context, ~5 seconds Service Info, ~15 seconds Product Interaction, ~5 seconds Results/CTA. In a 60-second ad, you might double those, but you'd still want to front-load the most critical info early on. Watch your edited video and note if there are any lulls where the energy drops – those are spots to tighten up. Perhaps the personal story is too drawn out – trim it to the essence. Or maybe the product demo lingers too long on one shot – cut sooner or add a secondary angle to maintain momentum. The goal is a **snappy, engaging rhythm** that **holds attention through each "beat" of the story**. If you've kept someone watching through 25 seconds, they're likely interested enough to finish the ad – but the first 5–10 seconds are where the majority will decide to stay or go, so make those count the most.

· **Visual and Audio Cues:** Use editing to your advantage by incorporating **visual cues** that guide the viewer's focus. We've mentioned on-screen text a few times – it's worth emphasizing: **subtitles and bold callouts can significantly boost engagement and comprehension**. Remember that up to *85% of online videos are watched on mute*. If someone doesn't hear the heartfelt story you're telling, they should at least *read* the gist of it. That's why adding captions for dialogue and big, punchy text for key points (problem, product name, result achieved, etc.) is almost standard practice in UGC ads. Many successful ads have the speaker's main points in large text at the top or middle of the screen in sync with the voice – this keeps even the silent scrollers in the loop. Also consider adding **graphic elements** like progress bars, check marks (✔) when a benefit is mentioned, or an outline effect to highlight the product in use. These should be used sparingly and purposefully, as *visual emphasis.* For audio, if the person speaking didn't explicitly say a strong hook line or CTA, you can use a bit of text-to-speech or a quick voiceover addendum for clarity (just ensure it matches the tone). And as noted earlier, background music or sound effects can help maintain a vibe, but **keep audio levels balanced** – dialogue should always be clearly audible over any music.

- **Pattern Interrupts:** To keep engagement high, especially in longer ads, introduce **pattern interrupts** – slight shifts that reset the viewer's attention clock. This could be a quick cut to a different camera angle, a sudden but relevant text animation, a zoom effect, or a short on-screen graphic (for example, a big "Before → After" divider appearing momentarily, or a meme-like reaction sticker popping up for humor if appropriate). UGC editing often borrows from trending social media styles, which include quick memes or visual jokes. If the tone of your ad allows, a tiny dose of humor or an unexpected visual can re-engage someone's interest. Example: In a skincare ad, right after showing the messy "before" face, you could flash an exaggerated emoji face for half a second – it's a fun blink-and-you-miss-it moment that can make the content feel more native to social feeds. Just be careful that these interrupts **don't distract from the core message** or make the ad feel disjointed. They should feel like natural parts of the persona's expression.

- **Consistency and Theming:** While you vary shots and interrupts to keep it lively, maintain a **consistent thread or theme** that ties it all together.

This could be the personality of the presenter (their unique humor or style carries through), a color scheme (using brand colors subtly in text or graphics), or a narrative device. For instance, some UGC ads use a **day-by-day vlog style**: "Day 1, trying the product... Day 3... Day 7... Day 30 result!" This inherently structures the sequence and gives viewers a clear path to follow. Another might use a **list or countdown**: "Top 3 things I loved about [Product]" – where personal context is #1 (relatable reason they needed it), service info and demo come as #2 (the experience of using it), and results as #3 (the payoff). In an advanced chapter like this, we encourage you to experiment with such creative structures *built on top of* the four core sections. They can differentiate your ad and make it more memorable, as long as the fundamental story still comes through.

Let's not forget testing and iteration: Even the pros rarely get a perfect sequence on the first edit. Part of advanced UGC editing is **analyzing viewer behavior and feedback**. Many editors produce a couple of variations – maybe one version jumps straight to a dramatic result before rewinding to context, while another sticks strictly chronological – and then they test which one retains viewers better. Pay attention to metrics like 3-second views, 10-second views, etc., if you have access to them, or simply get feedback from colleagues or the client. Does the story make sense? Is anything confusing or boring? The more you refine your sequencing craft, the better you'll get at intuitively knowing what works. But staying open to adjustments is key; sometimes a minor tweak like moving a single sentence earlier can boost retention significantly.

Before we wrap up, here's an example of a **well-sequenced 30-second UGC ad** outline to illustrate these principles in a cohesive flow:

- **0–3s (Hook & Personal Context):** "I used to have to choose between drinking coffee or dealing with migraines every afternoon..." (Speaker holds head, grimacing; text on screen highlights "migraines every afternoon?") – Viewers hooked by a common problem.

- **3–5s (Personal Context cont'd):** "...and as a software developer, I *live* on coffee." (Shows him at a desk with multiple empty cups – a touch of humor and relatability.)

- **5–8s (Transition to Service Info):** "I knew I had to find a better way to get through the day, and that's when I found **JavaZen**." (Shows him holding a product – perhaps a supplement drink or an alternative coffee; product logo appears.)

- **8–12s (Service Info):** "JavaZen is a natural energy drink with zero crash. It's basically jitter-free coffee in a can. And no more headaches! They even have a sampler pack to try." (Mix of him talking and quick cut to product shots; text callout: "No crash, No headache!" and "Sampler pack available").

- **12–20s (Product Interaction):** Montage – he opens a can of JavaZen, takes a sip. "First time I tried it, I was shocked – I felt alert *and* calm." (Shows him coding efficiently, smiling). Quick cuts: the can's label (highlight "All Natural"), him pointing to clock "4 PM – no migraine!" with a thumbs up. Perhaps even a side-by-side of him day before (tired at 4pm) vs after JavaZen (energized at 4pm). The energy is upbeat and visuals dynamic.

- **20–27s (Results):** "It's been 3 weeks on JavaZen. No more 3 o'clock crashes, and I've ditched coffee for good." (Shows him tossing coffee beans in trash playfully, or him jogging after work instead of napping). He looks genuinely happier and productive. Maybe overlay a quick quote from another user like a text message: "Dude, this stuff is a game changer – coworker".

- **27–30s (CTA):** "If you rely on coffee but hate the crash, you gotta try JavaZen. Trust me, it's a game-changer." (He raises the can as if in a toast; final screen shows product and a big "Try JavaZen ➜" text with the brand logo).

In this sequence, notice how each part flows logically, and at no point do we linger too long on one thing. There's context, solution intro, demonstration, and result, all within 30 seconds, told like a personal success story. That's the power of tight editing combined with a solid UGC structure.

Now, beyond technique, let's talk about content sourcing – because one hallmark of an advanced editor is the ability to find *great material* to work with, sometimes from the most unexpected places. This brings us to a short sidebar story that illustrates the creativity and resourcefulness that can set you apart in UGC editing.

Sidebar: Mining Gold from Unexpected Content – A Real-World Example

One of the most important skills you can develop as a UGC editor isn't just cutting footage — it's recognizing *gold* when you see it... even when it doesn't look like an ad.

Let me give you a real example from my own work.

I was working with a teeth whitening and oral health brand, and I had already gone through their usual content – testimonials, product shots, lifestyle footage. I needed something fresh, something with real authenticity. Someone from the brand sent me a few YouTube podcast interviews the CEO had done. They weren't "ads" in any traditional sense – just two-hour-long podcasts.

But I watched them all. I listened closely for emotional resonance, product credibility, and those magic unscripted lines that stick in your head.

Using the same four-part structure we've outlined in this chapter, I built a 40-second ad using only content from those podcasts. No b-roll. No product shots. Just subtitles and the CEO and the podcaster talking face to face.

Here's what made it work:

- I created **4 different openers**, each designed to stop the scroll. The winning opener started with the podcaster saying:

  "And this is no joke. I had gotten out of prison... and my teeth were like butter stick yellow."

It was bold. Scroll-stopping. And completely real.

141

From there, we moved into **the onramp**:

The podcaster shared his experience trying whitening strips —

> "They hurt my teeth..."

which set up the problem beautifully.

Then the **Service Info** hit naturally in the story – he says someone sent him the brand's product, he tried it, and saw results overnight. Smooth transition into Product Interaction and Results, all embedded in their unscripted conversation.

The CEO then started talking about *why* their product works – how it whitens teeth without causing pain. His passion and credibility came through completely naturally. Again — **no script, no staging, no visual fluff. Just conversation and subtitles.**

The result? That ad became the top-performing ad on the account for several months. We reused it across different promotions, cut variations for retargeting, and even posted it as organic content — where it became the most-viewed video on their entire account. Not only that, it boosted sales across the board because the ad was focused on the *brand*, not a single product. It was about trust.

One line from the ad:

> **"My teeth were butter stick yellow"** — became so iconic that I repurposed it as a **header** for other ads. That quote alone drove up retention and performance.

Final Takeaway: Always Be Mining for Hidden Gold

As we wrap up this advanced guide to UGC editing, remember: your best-performing ad might already be sitting in your client's drive or buried in a podcast no one thought to look at.

Whether it's a raw testimonial, a product review, a founder ranting on a livestream, or even just a customer email — **there is gold hidden everywhere**. Your job is to spot it, shape it, and build it into a compelling narrative.

The four-section structure we've covered (Personal Context → Service Info → Product Interaction → Results) gives you a reliable framework. You've learned how to pace your sequences, structure hooks, keep attention, and deliver a powerful CTA.

But **what will set you apart** is your ability to look beyond the obvious, get creative, and **trust your ear** for authentic moments.

Start by dissecting your top performers. Ask:

- What made this work?
- Was it the opener?
- The storytelling arc?
- A specific quote that felt too real to ignore?

Then reverse-engineer it and try those patterns in other edits. Test new openers. Test different orderings. Swap a scripted CTA with a casual line from a podcast. And most importantly: **always watch everything**. The next viral ad might be hiding in minute 46 of an unedited clip someone forgot to delete.

UGC editing isn't just a skill. It's a mindset. A scavenger hunt.

And when you treat it like that, you'll start to see stories — and results — where others just see raw footage.

So go mine your next piece of content gold.

11

Comedy, Tone & the Message-First Mindset

Humor is a powerful creative tool. It can stop a scroll, spark a share, and build a brand voice that feels human and relatable. But here's the hard truth that editors and creators must understand:

Funny doesn't always convert.

In fact, it *usually* doesn't — at least not on its own.

You might see a funny skit or clever punchline generate sky-high thumbstop rates, even great watch times. But that doesn't mean the ad is selling the product. Too often, comedy becomes a distraction from the core message. If viewers laugh and move on without remembering what the product was or what problem it solves, then the ad has failed to convert.

That doesn't mean comedy has no place in paid social — far from it. Some of the best ads use humor to amplify the story. But the key is understanding when and how to use it.

The Golden Rule: Message First, Comedy Second

Funny isn't a strategy. **Selling is.**

If your ad is built *only* around a joke, a skit, or a meme format without delivering a clear product benefit, you're essentially making entertainment — not advertising. And while there's overlap, the end goal is different.

> A joke can be a great *hook*, but **it's the *message* that converts.**

That means you can absolutely lead with something comedic — a goofy confession, a viral meme, a sketch-style bit — as long as you quickly pivot into a structure that educates, demonstrates, or sells the product.

Let's look at some examples of comedy **that works**:

- A creator opens with a sarcastic "day in my life" voiceover showing their chaos, then says: "Okay but real talk, this planner actually changed my mornings..." → and transitions into product interaction.
- A guy jokingly pretends he's married to his air purifier, then says "No really — this thing changed my sleep game..." → followed by clear features and results.
- A skincare ad opens with a parody of every influencer routine ("Step 87: Pour moon water into your socks") then cuts to the real pitch: "Forget all that — I just use this one product."

In all of these, humor is the hook, not the whole story. They still follow the onramp → service info → product interaction → result sequence.

When Humor Hurts Performance

There are patterns to watch out for when comedy flops:

- **The joke overshadows the product** – If viewers remember the punchline but not what's being sold, the ad has no stickiness.
- **Tone mismatch** – A product that helps with serious problems (health, finances, mental wellness) often doesn't benefit from humor unless it's delivered *very* thoughtfully.
- **Unclear CTA** – Funny ads sometimes forget the basics: what's the offer? what should I do next?

Worse, clients may fall in love with a "funny concept" without checking if it aligns with their goals. As an editor, your job is to protect the message. If you can make it funny *and* sell, amazing. But don't trade clarity for comedy.

Low-Fi or Polished? It Depends on Tone

Comedy also intersects with execution style. Some of the most effective funny ads are **low-fi, raw, and casual**. Why? Because they feel more authentic — like a friend joking on FaceTime.

Others, especially for larger brands, **lean into polished comedy**: think scripted skits, camera crews, lighting setups.

Both can work. The right choice depends on:

- **Brand identity** – Is the brand cheeky and irreverent? Or serious and trustworthy?

- **Target audience** – Gen Z often prefers casual humor. Older demos may respond better to polished delivery.
- **Channel** – TikTok leans raw. YouTube skippables might lean scripted.

The key is to stay *on-brand* and not fake the funny. If the creator or editor isn't naturally funny, forced comedy will feel... off.

Tone Supports Brand Identity

Beyond comedy, *tone* is your guiding north star. Everything from word choice to visuals to music informs how the ad makes the viewer feel.

Examples of tone:

- **Friendly and Casual** – "Hey guys, I gotta show you this thing real quick..."
- **Empowering and Confident** – "You deserve better hair. That's why I switched to [Brand]."
- **Emotional and Raw** – "After my diagnosis, I had no idea what to expect..."
- **Playful and Bold** – "I never thought I'd use the word sexy to describe a vacuum, but here we are."

You're not just cutting footage — you're shaping the *voice* of the brand. Keep tone consistent across the ad, and make sure it fits the product, the platform, and the people.

Key Takeaways: Comedy & Tone

- ✔ Comedy can *hook*, but must lead into a strong product message.
- ✘ Don't rely on laughs to do the selling — clarity beats cleverness.
- ◉ Always prioritize the CTA and value prop. Comedy should support, not distract.
- 🕹 Match the tone to your audience and platform.
- ↩ Test funny versions *against* straight-shooting versions. Let the data decide.
- 🎭 Keep a swipe file of funny ads that *also sell* — study what makes them work.

Message Over Mood

This chapter isn't anti-humor — it's about understanding its role. Funny can go viral. Funny can build trust. Funny can sell. But only when the message leads.

At the end of the day, a strong hook, a relatable story, a product people want, and a clear CTA will always outperform a comedy sketch with no point.

Make them laugh *and* make them convert — **that's the real win.**

12

Where to Watch Real Ad Examples

At this point, you might be wondering: *"I've read all this... but where can I actually see examples of the kinds of ads we're talking about?"*

The short answer: **every time you open social media.**

Whether you're scrolling through TikTok, Instagram, Facebook, or YouTube Shorts, you're being served performance-driven video ads. So next time you scroll—pay attention. Watch like an editor. Ask:

· What archetype is this ad using?
· Does it have a strong hook?
· How is it structured?
· What would I do differently?

Every ad in your feed is a free case study. And once you know what to look for, you'll start spotting trends, patterns, and performance signals everywhere.

Want to go deeper? Use Facebook Ads Library. It's a public tool that lets you search and watch current ads running from almost any brand in any country. Just type in a company name or product category and you'll unlock a goldmine of real creative examples to study and reverse-engineer.

III

Part III: Experimentation, Business & Scaling

Test like a scientist. Build your portfolio. Choose your path: grow in-house, freelance, or launch your own agency.

13

Building Your Portfolio (Even with No Clients)

If you're reading this chapter, you're probably asking the question every new editor has asked at some point: "How do I build a portfolio if no one will hire me without one?"

Let's be blunt — the best way to build a strong, fast-moving portfolio as a paid social video editor is to get hired by an agency.

The Agency Route: The Fast Track

Working at a creative agency (even as a freelancer or part-timer) exposes you to multiple brands, multiple editing styles, and rapid feedback loops. You'll crank out dozens — sometimes hundreds — of videos per month. And while you don't own the work (it's not "your brand"), you absolutely **can and should** include the best-performing or most creative ads in your portfolio, as long as you clarify your role in the production.

Agencies are fast-paced, but they teach you how to think like a performance editor. You'll work under creative producers, get access to briefs and brand guidelines, and learn what works — and what gets rejected. It's bootcamp and college rolled into one.

The In-House Brand Route: Great Guidance, Narrow Scope

The second best way? Work directly for a brand that has a solid internal creative team. If you're paired with strong producers or a head of creative who knows paid social, you'll learn how to structure edits properly and how to analyze performance. The downside is that you're working with only one brand. So while your skills grow, your portfolio won't show variety. That's okay at first — you can build your versatility later.

But What If You Can't Get Hired Yet?

Then you create. Simple as that.

You don't need permission to start editing.

If no one's hiring you yet, make content anyway. Spec ads. Remixes. UGC-style mockups. Take raw content from brands you love (many creators post uncut versions), and build your own ad versions using the frameworks in this book.

Don't wait to get paid to start producing. Every ad you edit — even on your own — sharpens your instinct, teaches you pacing, structure, and tone.

Make 3–5 spec edits and drop them in a simple portfolio folder, Google Drive link, or Notion board. That's **enough** to get your foot in the door somewhere.

My Path: From Intern to U.S. Paid Social Editor

When I arrived in the U.S., I barely spoke English. I didn't know how to communicate with companies — or even people. But **I did know how to edit**. I had taught myself multiple editing programs, and I used that skill to land an internship at an advertising company doing basic content:

- Lawyer ads
- Celebrity news recaps
- General YouTube/social media edits

It wasn't glamorous — but it gave me a start. That internship became a job, and during that time, I also made frequent trips to Mexico to meet people and find new opportunities. I didn't know anyone there either — I'm from Venezuela — but I knew I could connect better in Spanish, and that gave me an edge.

Eventually, I landed commercial editing gigs for brands like H&M, Nike, and Versace. These weren't paid social ads — **but they looked great.** And they opened the door for me to get paid social work in the U.S. Once I had that, I started building my *real* portfolio.

The truth is: you don't need the perfect job, the perfect contacts, or the perfect plan. You need action. You need output. **The more you edit, the more you attract opportunities.**

Your Portfolio Game Plan

Here's how to actually build your portfolio, step-by-step:

1. Choose Your Starting Point

- If you're just beginning, create 3–5 short ads from scratch using raw footage or public assets.
- If you've done any freelance or personal work, reformat it into ad-style content.

2. Use the Structures From This Book

- Don't just cut randomly. Use the four-section structure from Chapter 10.
- Make sure your edits **sell** something — not just look pretty.

3. Package Smartly

- Upload your ads to a Google Drive folder or create a simple Notion portfolio.
- Title each piece with the product name and what it demonstrates (e.g., "Skincare UGC Ad – Spec | Retention Structure").
- Include a brief description: what type of ad it is, what structure it follows, and your role.

4. Share Relentlessly

- Post on LinkedIn or Instagram. Tag editors or creatives you admire.
- Apply to internships or junior editor positions with your link.
- DM brands you like and say, "Here's an ad I made using your content — happy to do more."

Bonus Tips

- **Study performance ads daily**: Save the ones that stop you mid-scroll. Try to reverse-engineer them.
- **Track what you make**: Keep a spreadsheet or Notion board to log your edits and what skills you learned from each.
- **Collaborate with creators**: Offer to edit their UGC in exchange for using

it in your reel.

Final Word

The single best advice? **Create content.**

Whether it's spec work, past client projects, or re-edits — you need proof of your skills. Not theories. Not potential. **Proof**.

This industry moves fast, and the editors who show work — not just talk about it — are the ones who get noticed.

Create. Share. Improve. Repeat. That's how you build your portfolio. That's how you get hired.

And most importantly, **that's how you get good.**

14

From Freelancer to Creative Pro

There's a moment every editor hits — you're no longer just someone who knows how to cut a video. You've delivered results, built creative instincts, and maybe even landed a few recurring clients. So now what? This chapter is about **evolving your mindset** from "just an editor" to a creative pro that brands and agencies can't afford to lose.

What "Creative Pro" Really Means

You're not just pressing buttons anymore — you're solving business problems with creativity. Being a creative pro means:

- Thinking in **results**, not just edits.
- Understanding how to **communicate your value**.
- Making yourself **indispensable** in the process — not just someone they can swap out for cheaper.

It also means you're no longer just taking orders — you're contributing to ideas, spotting what works, and helping shape the content that drives performance.

Positioning Yourself Like a Pro

Most freelancers stay stuck because they never learn to **market themselves**. That doesn't mean being loud or fake — it means showing people that you're legit and that you know how to **drive results**.

Here's how to position yourself for higher-paying, higher-trust opportunities:

1. Be Easy to Work With

The best editors aren't just skilled — they're low-maintenance and reliable. Reply to messages fast. Hit deadlines. Don't take feedback personally. If someone likes working with you, they'll keep hiring you.

> **Pro tip:** *Clients remember how easy or hard it was to get things done with you — not just how slick the final edit looked.*

2. Show Results, Not Just Reels

Your portfolio should go beyond "cool edits." Show **what the video achieved** — high engagement, great comments, conversion stats (if you have them), or even just social proof like "client said this ad got more sales than anything else last month."

Add context to your work. Frame it like this:

> "Edited this ad for X brand. Hook retention was 42%. ROAS was 2.8x. It ended up running for 3 months straight."

Even without metrics, a quick caption like "ran as the main promo video for Black Friday" tells people you're doing real work that matters.

3. Have a Professional Presence

You don't need a full-blown website, but at minimum you need:

- A clean Notion board or PDF portfolio
- A solid LinkedIn profile that shows who you've worked with
- A quick intro pitch that explains what kind of editor you are

> "I specialize in high-retention UGC ads for paid social — brands usually bring me on when they need fast-turnaround edits that still convert."

That's short, clear, and shows you know your space.

How to Pitch Clients (and Get Paid)

1. Start with Value

Don't just ask "do you need an editor?" Say:

> "Hey, I saw your brand running UGC ads — I specialize in turning raw content into scroll-stopping paid social videos. Would love to help you test some new creatives."

2. Have a Simple Offer

Make it easy for them to say yes. Something like:

> "I usually work on a flat-rate per ad (e.g., $250/video for 30–60s). I can also do batch rates if you're testing multiple hooks."

Always be open to retainer models if you want stability.

> *Note: If you're just starting, don't undersell yourself too much. You can offer lower rates at first, but set a clear path to raise them. If your work converts, your value multiplies.*

3. Get Clear on Scope

This avoids endless revisions and unclear expectations. Set the rules early:

- What you're delivering (e.g., 1x 30s ad + 3 cutdowns)
- How many rounds of revisions
- Turnaround time
- Payment terms (e.g., 50% upfront, 50% on delivery)

Use simple contracts — even a one-pager with a clear email agreement can work.

Building Referral Systems

1. Deliver Like a Beast

If you do amazing work, clients **will** refer you. But sometimes you have to ask:

> "Glad you liked the ad — if you know any other brands looking for performance video, feel free to send them my way."

Most people are happy to help, but you need to plant the seed.

2. Tap Old Clients

Every 3–6 months, reach out with a message like:

> "Hey! Just checking in — would love to help with any new creative tests you've got coming up."

Or share a recent win:

> "Just helped another brand scale a hook variation that crushed — happy to help you build something new too."

3. Network Smart

Get in relevant communities (Discord groups, X, LinkedIn). Don't just lurk — contribute. Share what you're learning about editing, testing, performance.

If you become known as someone who knows how to make ads that convert, work will come to you.

You're Not Just a Freelancer Anymore

This chapter is about a mindset shift. You're not waiting for someone to give you work. You're actively building a reputation as a **creative operator** — someone who solves real business problems with high-performing content.

The difference between a freelancer and a creative pro isn't skill — it's **how you see your role**.

Clients don't want just an editor.

They want someone who:

- ✓ Understands performance
- ✓ Works fast
- ✓ Takes feedback
- ✓ Makes smart decisions
- ✓ Wants the brand to win

If you can be that person, you'll never run out of work.

15

Beyond the Timeline – Growing Into a Creative Collaborator

If you're reading this chapter, chances are you've already built some editing chops. You can cut clean. You know what a hook is. You understand structure, platform nuances, and performance metrics. That's good. But to become indispensable—to go from "someone we hire" to "someone we need"—you've got to go beyond the timeline.

Editing is only one part of the creative process. The more value you bring outside of cutting footage, the more opportunities open up. In this chapter, we'll talk about how editors become collaborators, what roles you'll interact with, and how your work intersects with theirs. We'll also cover the extra tasks you might be asked to take on—and how mastering them can grow your career.

The Creative Ecosystem Around a Paid Social Editor

Even if you're working solo on the timeline, you're rarely creating in a vacuum. Here are the most common collaborators you'll work with in the paid social space—and what they'll expect from you:

- **Creative Director:** Owns the brand's high-level creative vision. They're not in the trenches cutting ads, but they often sign off on ideas and tone. They care if the creative is on-brand and consistent.

- **Creative Producer:** This is your project manager and creative partner. They'll send the briefs, track deadlines, gather raw assets, and help prioritize deliverables. Good producers are worth their weight in gold— collaborate with them early and often.

- **Ad Manager / Performance Marketer:** They speak in metrics. These are the people watching thumbstop rates, ROAS, and click-throughs. They'll often tell you what's working (and what's not) and expect you to take that into account. They don't care about how "cool" a transition is—only whether it converts.

- **Brand Designer:** They maintain the brand's visual identity. They may give you style guides, logo lockups, font specs, and color palettes. Respect their work—it makes your work look better too.

- **Graphic Designer / Animator:** These folks can elevate your edits with polished motion graphics, captions, or layout design. If you don't have animation skills, become great at collaborating with those who do.

> **Pro Tip:** Learn to speak their language. Ask "What's the offer we're pushing?" or "What's our thumbstop goal?" When you talk in terms of goals, data, and brand values—not just video tricks—you become a trusted partner, not just a pair of hands.

Other Tasks You May Be Asked to Do

Once clients or agencies trust your edits, they may start looping you into other tasks. Here are some common ones:

1. Creating Briefs

Sometimes you'll get a fully fleshed-out concept. Other times, you'll need to help shape it. You might be asked to write a brief that outlines the story you're cutting: the hook idea, selling points, visual style, and CTA. If you've already been reviewing metrics and top performers, you're in a great position to guide this.

2. Writing Scripts for Talent

In UGC or testimonial-style ads, brands often need scripts that feel casual but still hit key selling points. You might be asked to write or edit these scripts. They usually run 30–60 seconds and follow the same structure we've discussed: hook → context → product → results → CTA. If you know what converts, this becomes second nature.

3. Designing Statics or Thumbnails

This one surprises a lot of editors. But yes—if you can use Canva, Figma, or Photoshop, you might be asked to whip up simple thumbnails, stills, or ad statics. Often it's just about repurposing headlines from your video into scroll-stopping graphics.

4. Suggesting New Concepts

This is where the real leap happens. If you consistently deliver results, your client or agency might start asking: "What should we test next?" They'll expect ideas for fresh angles, new formats, and different archetypes. You're no longer just executing—you're now a creative driver.

How This Makes You More Valuable

Agencies and brands are hungry for creatives who can wear multiple hats. If you can:

- Cut converting ads,
- Suggest testable concepts,
- Translate data into creative direction,
- And collaborate seamlessly with others on the team...

You become much harder to replace.

In a world where editors are often commoditized, being a **creative collaborator** sets you apart. You're not just reacting to briefs—you're helping build the creative roadmap.

Recap: How to Level Up From Editor to Collaborator

	Skill/Action	Why It Matters
1	Learn the language of performance (CTRs, ROAS, etc.)	Builds trust with Ad Managers & Producers
2	Collaborate proactively with Creative Producers	Makes your life easier + keeps projects on track
3	Contribute ideas, not just execution	Shows leadership and creative thinking
4	Help shape briefs or scripts	Elevates your role from post to pre-production
5	Learn basic graphic tools (Canva, Figma, etc.)	Increases your value across multiple formats

Final Thought

The further you go in this career, the more you'll realize that editing is just one tool in your belt. The editors who thrive long-term are the ones who understand the full creative ecosystem—and know how to move through it with confidence, clarity, and collaboration.

You don't have to do everything at once. But if you start stepping outside the timeline a little more each week, soon you'll be sitting at the creative table—helping shape the ads, not just cut them.

16

Experiment Like a Scientist, Edit Like a Creative

The best paid social editors aren't just creative — they're strategic. They're not just making videos that *look* good; they're creating videos that *work*. That means testing. That means iterating. That means learning from what the data tells you, and then building something better.

In this chapter, we're going to explore the mindset that separates good editors from elite ones: the ability to edit like a creative *and* think like a scientist.

There Is No Perfect Formula

No matter how much experience you have, no matter how polished your cut is, there is no guaranteed "perfect edit." Even the smartest hook, the most polished sequence, or the funniest line can fall flat in the feed. Why? Because we're editing for real people — with real moods, scrolling habits, and unpredictable behaviors.

That's why testing isn't just helpful — it's *necessary*.

The best editors know that every video is a hypothesis. Every variation is a data point. Every round of performance feedback is a clue about what *might* be working — and what definitely isn't.

> "Great editing isn't about getting it perfect the first time. It's about knowing how to learn from each version and get closer to what converts."

Creative Meets Scientific: The Dual Mindset

You need to balance two worlds:

- **The Creative Mindset** helps you explore, take risks, and find fresh angles.
- **The Scientific Mindset** helps you test those ideas and figure out what's *actually* working.

The magic happens when you can toggle between both.

Maybe your gut tells you this quirky, sarcastic line will hit — great. Try it. But also test a version with a more earnest, relatable tone. Look at the data, see what performed better, and *keep building.*

Top editors don't argue with the results — they *use* them.

- Launch date
- Key metrics (3-sec view rate, hold rate, CTR, CVR)
- Notes, learnings, or insights

Pro tip: Always log your hypothesis *before* testing. What do you think will happen? That way, you're practicing intentional creative thinking — not just guessing in hindsight.

Let the Data Guide You — But Not Define You

Yes, numbers are crucial. But they're not the whole story. Sometimes your best-performing ad is one you *almost didn't publish* — because it broke all the "rules."

This is why experimentation matters so much. You're not testing to *follow the formula*. You're testing to *find* the formula that works for that brand, that audience, that offer, right now.

Let the data be a compass, not a cage.

Don't ignore patterns. But don't let them stop you from trying something new.

Real Experiment Examples: Turning Flops into Top Performers

Let's get practical. Here are real examples of small creative tweaks that turned a "meh" ad into a top performer:

1. The Hook Switch

- **Original:** "Here's how I stay organized every morning."
- **Tested:** "I used to forget everything — until I found this."
- **Result:** 3-second view rate jumped by 45%. Same video, same script, just a better opener.

2. UGC vs. Brand Voice

- **Original:** Voiceover by brand spokesperson.
- **Tested:** UGC testimonial with captions only.
- **Result:** UGC version had 3x higher retention and 2x better CTR — likely due to authenticity.

3. Adding Social Proof On-Screen

- **Original:** Product demo only.
- **Tested:** Same demo + overlay text: "Over 30,000 5-star reviews."
- **Result:** Conversion rate improved by 28%.

4. "Bad" Footage > Polished Footage

- **Original:** Cinematic lifestyle b-roll with voiceover.
- **Tested:** Raw, shaky front-facing selfie of user saying "I can't believe this worked."
- **Result:** Raw version outperformed the cinematic version across every metric.

5. The "Silent Scroller" Fix

- **Original:** No subtitles.
- **Tested:** Exact same video with captions added.
- **Result:** +40% watch time, especially on mobile.

6. Talent Swap

- **Original:** Sales sequence performed by Creator A.
- **Tested:** Same exact script and b-roll with Creator B.
- **Result:** Creator B version outperformed by 60% CTR. Why? Hard to say. Better tone? More relatability? Just worked. That's the magic of testing.

Final Thought: You're Not Just an Editor — You're a Creative Scientist

The best editors don't wait to be told what works — they go find out. You're not just making videos, you're *running experiments that sell.* Every variation is a chance to learn. Every test is an opportunity to improve.

So embrace it.

Test fearlessly. Track obsessively. Learn constantly. And when you find that magic combo of hook + context + demo + result?

Double down. Then test it again.

Because what works today might flop tomorrow — and what flopped last month could be your next big win with just one small tweak.

17

Scaling as a Creative Business

Once you've mastered editing for paid social, the next natural step is growth. For some, that means becoming a lead editor at an agency or brand. For others, it means building a client list and turning freelance editing into a scalable business. There's no single path forward – but understanding the landscape can help you choose which direction fits your goals best.

Working with Agencies vs. Brands

One of the most important decisions you'll make as a professional editor is whether to work with agencies or directly with brands. Each has its own structure, expectations, and pros and cons.

Agencies: High Volume, High Structure

Agencies are machines. They operate with built-in systems and layers of support. You'll often be given everything you need to succeed:

- A creative producer guiding your process
- A content brief already written

- Pre-selected b-roll and approved talent
- Clear structure and naming conventions
- Pre-determined deadlines and ad variations

You'll likely work across multiple brands every week, which is a great way to build a varied portfolio fast. It's also a crash course in how different industries sell – fashion, supplements, tech, personal care, fitness – and you'll learn to edit across tones and formats.

The downside? Agencies often run fast and lean. You may be juggling five or more projects at once. Deadlines can be tight, and client feedback can change quickly. The work is rewarding, but it requires discipline, communication, and adaptability. That said, agencies tend to be more stable than internal brand teams – layoffs are less frequent unless the agency itself is restructuring.

Brands: Focused Growth, But Less Structure

Brands are a different story. Typically, they hire in-house editors because they're trying to produce ads faster, cheaper, and at higher volume than what an agency provides.

If you work in-house at a brand:

- You'll get to know the brand identity deeply
- You'll usually focus on just one product line (or a few)
- You may work closely with marketing, creative, and even product teams
- You're expected to deliver results that justify your cost (often trying to beat what the agency delivered before)

Many brand teams are lean. You might be one of only a few editors. You'll need to wear multiple hats, move fast, and sometimes build systems from scratch. It can be exciting – but also volatile. If the company cuts budget or changes strategy, in-house content teams are often the first to go.

Becoming a Creative Strategist: Making the Jump from Editor

If you've been editing ads for a while, you'll start to develop an instinct for what works and what doesn't. You'll notice patterns in performance, start crafting hooks proactively, and suggest b-roll that enhances a message. When that shift happens, you're stepping into strategy.

A creative strategist doesn't just execute – they plan. They build scripts, structure variations, and guide other editors. They interface directly with brands and producers to shape the creative approach. That makes you more valuable and opens new career doors.

If you want to make that jump:

- Start documenting what performs and why
- Build case studies or video breakdowns
- Offer to test your own scripts or structures
- Collaborate more closely with creative producers and ad managers
- Practice writing briefs and concept decks – even just for yourself

Editors who think like strategists eventually get hired as them. You don't need a new job title to start acting like one.

Growing into a Full-Time Career vs. Launching a Micro-Agency

You can build a strong career as a paid social editor without ever starting a business. Many editors go full-time with brands or agencies, rise to lead editor positions, and earn solid salaries. Here's what that path looks like:

Full-Time Editor – The Career Path

- Great benefits: health insurance, PTO, remote setup, tech support
- Creative stability and mentorship
- You get to focus on your craft without chasing clients
- Top earning potential is around $100K–$250K depending on company and seniority

Most editors don't break the $100K mark. Those who do are often lead editors or creative ops leads with added responsibilities. It's a strong career path with a lot of lifestyle balance – but the financial ceiling is real unless you grow into executive or cross-functional roles.

Freelance / Micro-Agency – The Business Path

- Unlimited earning potential
- Total ownership of time, clients, and rates
- No ceiling – but also no safety net
- You handle everything: contracts, invoices, software, taxes, client relationships
- You rely on your own resourcefulness and network to grow

Starting a micro-agency is hard. Most of your early clients will come from referrals. That's why reputation and reliability are everything. If you're known as someone who delivers high-performing creative on time, you'll get recommended. Once that happens, you can raise rates and start hiring others to handle parts of the workload.

The downside? You're responsible for:

- Your own health insurance
- Buying or licensing music, footage, tools
- Managing timelines and expectations
- Taking no income when you take time off

That said, when you lose a client, you still have others. You're not wiped out. And when things go well, you can scale far beyond what a full-time job would ever offer.

Which Path Should You Take?

This comes down to your ambition, your lifestyle goals, and your risk tolerance.

Do you value security and structure? Then aim for an in-house or agency role and grow your title over time. Want full control, ownership, and no earning cap? Start building your own freelance business or creative shop.

Neither path is easy. Neither is "right" or "wrong." You'll face challenges in both. You can lose a client. You can be laid off. You can have slow months. That's just the nature of creative work.

What matters is building skills, creating value, and putting yourself in motion. Keep showing up. Keep experimenting. Keep improving your craft. The path becomes clear when you start walking it.

18

Owning Your Career, Craft & Future

You've made it to the end — and by now, you know this book wasn't just about editing video. It's about owning your career in a rapidly changing industry where creative meets performance and where the ones who win aren't just the best editors — they're the ones who know how to adapt, test, sell, and scale.

You Don't Need to Be a Master Editor

You need to be a performance-minded editor.

You learned early on that clean transitions, fancy effects, and expensive gear don't win in paid social. What wins is clarity. What wins is messaging. What wins is structure, strategy, and empathy for the viewer.

Your edit is a message. Your cut is a decision. Your timeline is a sales tool.

Scroll Psychology > Polish

The scroll is ruthless. You have 1 second. Maybe less.

Understanding archetypes, emotional triggers, and the science of thumb-stopping content gave you the blueprint to start strong and finish with intent. Every chapter after that drilled this home: hook early, sell clearly, prove value, and structure for results.

Build Like a Pro

From job folders to sequence naming, your workflow now mirrors what top agencies use. Not because it looks good — but because it keeps you moving fast and scaling efficiently. You're now set up to iterate, test, and track like a real creative operator.

Metrics Guide, Data Decides

You've seen how CTR, CVR, CPA, and ROAS are not just numbers — they're signals. They tell you what works, where it's breaking, and how to fix it. Now you know how to read them and use them to make your next edit better than your last.

You're Not Just an Editor — You're a Creative Producer

The middle of the book showed you how to pull apart a top performer, test versions, and think like a strategist. And that's what this whole space rewards — people who don't just cut clips, but build ideas, pitch angles, remix scripts, and chase performance like scientists.

This is your edge. Own it.

What Now?

Here's where the paths split. But it's all in your hands.

Path 1: The Career Climb

- You can go work for agencies. Learn the ropes.
- Grow from editor to lead editor to creative director.
- Get great benefits, mentorship, and the safety of a team.
- You'll ship work every week and be a part of big performance wins.
- It's a career with a ceiling — but it's also a great life.

Path 2: The Independent Route

- Or you go solo. Freelance. Build your own micro-agency.
- You'll deal with clients directly, pitch, strategize, experiment.
- There's no ceiling — just your time, your rates, and your ability to scale.
- It's riskier. But it's yours.

> 💬 **Note:** You can always switch between these paths.
> Start in an agency, build your chops, then go solo.
> Or freelance for a while and land a great full-time role.

There's no one-way door here — your journey is flexible and evolves with your skills, goals, and lifestyle.

And there's no "right" path. Just the one that matches your ambition.

Final Lessons to Carry With You:

- The best editors don't wait for the perfect brief — they make something great with what they've got.
- Funny doesn't always convert. But curiosity does. So does emotion. So does clarity.
- There is no perfect formula. Only structured experimentation.
- Rewatch your own top performers. Dissect them. Steal from yourself.

- Watch your numbers. Track your versions. Deliver clean.
- Talent matters. But consistency matters more.
- If you can build value and solve problems — you'll always get hired.

This book isn't your trophy. **It's your toolkit.**

It's not meant to sit on a shelf or your desktop. It's meant to be used. Reread the chapters when you get stuck. Go back to the naming structure when things get messy. Refer to the metrics when a client asks what to test next. And when you hit a big win? Document it. Reverse-engineer it. Learn from it.

Because this space doesn't slow down.

Scroll speeds will get faster. Platforms will change. Trends will burn out. But the foundations in this book — clarity, story, testing, structure — will keep you grounded, relevant, and in demand.

So take what you've learned. Apply it. Experiment with it. Break it. Rebuild it.

And most importantly: **keep editing, keep converting, and keep growing.**

Why Hook. Edit. Convert.?

The title of this book isn't just a catchy phrase — it's the **mindset** of a modern creative editor:

- **Hook:** Capture attention in the first second. Make someone stop scrolling.
- **Edit:** Build a compelling story that informs, engages, and sells.
- **Convert:** Guide the viewer to take action — whether that's clicking, buying, or remembering the brand.

This is the **core loop of performance creative.** Everything else is refinement.

When in doubt, come back to it:

Did you hook them? Did your edit deliver? Did it convert?

That's your job.

That's your craft.

That's your edge.

This is just the beginning.

19

Bonus: The AI Advantage – A Modern Editor's New Toolbox

Your New Creative Assistant

AI isn't the future — it's here. And as a paid social video editor, it's one of the best tools you can add to your workflow. This chapter is your practical guide to using AI like a smart assistant — not to replace your creativity, but to help you move faster, think clearer, and get better results.

This section is a snapshot of the AI landscape as of 2025 — it's moving fast, so use this as a starting point, not the finish line.

In 2024, over 40% of small businesses started using AI tools — and marketers across the board are using it to draft scripts, analyze performance, and brainstorm creative faster than ever.

As a creative producer, you can use AI to:

- Analyze ad metrics like CTR, CVR, and ROAS in seconds
- Generate headlines, hooks, and ad concepts on demand

- Improve your scripts or refine messaging with brand-safe tone
- Summarize A/B test data and even suggest new test angles
- Speed up ideation without staring at a blank page

The best part? It's like having a junior editor or strategist working 24/7. You stay in control — but with more momentum behind every task.

Generating & Refining Scripts with AI

Faster First Drafts, Smarter Revisions

AI can help you write full ad scripts — not by taking over, but by giving you solid starting points so you're not staring at a blank doc.

Writing Scripts

You can prompt AI with:

> "Write a 30-second video ad for a skincare serum that clears acne. Start with a 3-second hook, explain the problem, show a testimonial, and end with a call-to-action. Keep the tone friendly and confident."

And just like that, you get a full draft with structure you can tweak — from the hook to the CTA. You stay in control, but now you're editing, not starting from scratch.

Editing and Polishing

Already have a script? Ask AI to:

- Make it snappier
- Cut it down to 20 seconds
- Match a specific tone (funny, bold, emotional)

You can also use templates: give the AI a structure (like a testimonial format) and ask it to plug in new product info.

Summarizing Long Testimonials

Got a long customer video? Ask AI to pull the best quotes or summarize the highlights. You'll get usable bites for subtitles or voiceover lines in seconds.

Important Note:

> AI helps draft — **you still finish**. Always review for brand voice, factual accuracy, and tone. Use it like a smart collaborator who writes fast but needs your eye to polish.

Summarizing Data & Writing Reports with AI

Turn Notes and Numbers into Insights, Fast

As a paid social video editor, you'll often need to share learnings from your ads — whether it's for a client, your boss, or just your own process. AI can help you translate raw results into clear, well-worded takeaways.

Quick Campaign Summaries

Let's say you have this:

Ad A: CTR 1.2%, CVR 10%, ROAS 2.0

Ad B: CTR 0.8%, CVR 15%, ROAS 2.5

You can ask AI:

"Write a quick insight explaining why Ad B performed better."

And it might give you:

"Ad B had fewer clicks, but its higher CVR led to a stronger ROAS, suggesting it attracted more qualified users who were ready to buy."

Summarizing Comments or Feedback

Got a bunch of viewer comments or client notes? Paste them in and ask AI to find the themes. Example:

"Summarize the top themes from 50 customer comments."

AI might say:

"Most viewers loved the humor and related to the pain point. Some wanted the product name introduced earlier. Overall sentiment: positive."

Super helpful for making sense of qualitative feedback fast.

Drafting Strategy or Recap Docs

> You can even use AI to help write testing recaps or strategy docs — just feed it bullet points or old reports and ask for a draft in the same style. It won't be perfect, but you'll have a clear head start.

AI as a Structural and Writing Assistant

AI isn't just useful for writing scripts — it can help you **structure and organize** your ideas too. Think of it as a creative partner that helps you figure out what to say and how to say it.

Structuring Your Script:

Let's say you have a bunch of selling points. You can ask AI:

> "What's the best order to present these in a 30-second video?"

It might suggest:

1. Hook (biggest pain point)
2. Product as the solution
3. Unique feature
4. Testimonial
5. CTA with urgency

Now you have a clear structure to plug your content into — perfect if you're stuck or want to double-check your instincts.

Simplifying the Message:

If your value prop sounds too complex, you can ask:

> "Explain this in one simple sentence at a 5th-grade level."

AI will help make your messaging **clear and easy to understand** — a must for fast-moving social feeds.

Fixing the Tone:

Not sure your copy hits the right vibe? Use AI to adjust it.

- "Make this script more exciting, but not pushy."
- "Tone this down to sound more professional."
- "Make it sound playful but still credible."

AI will rewrite your text with different tones so you can pick the best version. It's like having a built-in copy editor.

About Brand Voice:

AI doesn't know your brand's unique voice unless you **teach it**. If your tone is quirky or niche, you'll need to give examples. Still, AI can help get you 80% of the way there — then you fine-tune.

AI for Visuals & Voice – Great for Concepting, Not Yet for Final Ads

AI isn't just for writing — it can help with visuals and voice too. While the tech isn't quite ready for full-production ad assets, it's great for brainstorming, prototyping, and getting early ideas off the ground.

Visuals: AI as a Concepting Tool

Tools like **Midjourney, DALL·E 3 and ChatGPT** can generate high-quality images from text prompts. This makes them great for:

- Storyboarding scenes
- Building quick moodboards
- Mocking up ad ideas or layouts

Instead of hunting through stock images, you can prompt:

> "A woman jogging at sunrise in a futuristic city" — and get instant visuals to help you plan.

Caution: AI images come with legal gray areas. You often don't own the rights to use them commercially, and some may resemble copyrighted content. So for inspiration only – not final ads unless you're certain about the rights.

Video: AI for B-Roll and Transitions

Tools like **Runway ML Gen-2** or **Pika Labs** can create 3–4 second video clips from prompts. These are still rough, but useful for:

- Placeholder b-roll
- Testing concepts
- Creating abstract animations

They aren't ready for polished ads, but they can help you try new ideas quickly.

Voice: AI Narration When You Need It Fast

AI voice tools like **ElevenLabs** can generate super realistic voiceovers in seconds. Great for:

- Temp voice tracks for drafts or client pitches
- Testing different tones (youthful, serious, regional accents)
- Creating Voice Overs (VO)

Just type the script, pick a voice, and you're done.

You can even **clone a real voice** (with permission) for consistency in a series — like keeping the same voice actor tone across ads.

Be careful:

- Never clone voices without consent
- AI voices may lack emotion — humans are still better for nuance
- Check your usage rights if you plan to publish with an AI voice

Use AI visuals and voices to save time and build prototypes faster. Treat them like a creative intern — not a full-time replacement. You still run the show.

AI Tool Recommendations for Video Ad Creators

AI can be a massive time-saver — if you know the right tools to use. Here's a breakdown of the most useful AI tools for creative producers/editors, organized by function.

Text-Based AI Assistants

1. ChatGPT (OpenAI): The all-rounder. Great for brainstorming hooks, refining scripts, summarizing feedback, and even analyzing performance data.

- **Best for:** Rapid ideation, writing variations, script punch-ups, and light data insights.
- **Tips:** Give it context and direction — e.g., "Act like a copywriter. Write 3 TikTok hooks for a skincare brand targeting teens."
- **Caution:** Double-check any product facts or claims — it may "hallucinate" details.

2. Claude (Anthropic): Similar to ChatGPT but excels with longer inputs. You can feed it full transcripts, spreadsheets, or PDFs.

- **Best for:** Analyzing large documents (e.g. customer reviews or campaign notes), summaries, structured writing.
- **Tone:** More literal than ChatGPT — better when accuracy > creativity.

3. Jasper.ai: Built for marketers. Offers templates for ad copy, Facebook headlines, and more.

- **Best for:** Users who want pre-set workflows (great if you don't like writing your own prompts).
- **Bonus:** You can teach Jasper your brand voice by feeding it past copy.

Note: It's paid, but some teams love its organization and brand-style support.

AI for Visuals

1. Midjourney: Turns prompts into high-quality images.

- **Best for:** Concept art, moodboards, or storyboards.
- **Example:** "A skincare product on a glowing bathroom counter at sunrise."
- **Caution:** Don't use raw outputs in final ads — there are copyright risks. Great for internal mockups or pitching.

2. Runway ML (Gen-2): Creates short video clips (3–4 sec) from prompts.

- **Best for:** Prototype b-roll, abstract transitions, or testing visual styles.
- **Also includes:** Background removal, video upscaling, and text-to-video features.
- **Limitation:** Low-res, short length, and quality is hit-or-miss — not ideal for final output yet.

3. Pika Labs: Like Runway, but better with human realism and consistency in some cases.

- **Best for:** Motion mockups, animation ideas, product fly-throughs.

Note: Great for drafts — still evolving for polished ad use.

AI Voice Synthesis

ElevenLabs: Turns text into natural-sounding speech in seconds.

- **Best for:** Temp voiceovers in edits, testing tone (teen voice vs. deep announcer), localization demos (multi-language VO)
- **Advanced:** Clone a voice (with permission) for consistent use across ads.
- **Warning:** Never clone real voices without consent. Avoid misleading usage (e.g. fake testimonials).

AI for Reporting

Looker Studio (with AI summaries): Turns your performance dashboards into readable insights.

- **Best for:** Weekly ad performance recaps, auto-writing top insights (e.g., "CTR rose 20% after new hook")
- **Pair it with:** Google Sheets + ChatGPT integrations for deeper commentary.
- **Watch out:** AI might misread trends — always verify before sharing.

Pros and Cons of Using AI in Video Ad Creation

AI is changing the game for creative editors. But while it can boost your workflow, it comes with some challenges. Here's what you need to know.

Pros: Why AI Is Your New Creative Wingman

Speed & Efficiency: AI handles tasks in seconds that might take hours — script drafts, VO temp tracks, hook ideas. Faster cycles = more time to experiment and hit deadlines.

Rapid Iteration: Need 10 ad angles in different tones? Done. AI helps you test faster and scale more creative variations — essential for paid social where testing is king.

Unblocks Creativity: Stuck on a blank timeline or doc? AI gives you ideas to riff on — even if 9 fail, 1 might spark the winning concept. It's great for breaking creative ruts.

Consistency & Memory: AI can stick to a tone or repeat product details across outputs. You can train it to mention key points like "30-day guarantee" or match a persona throughout a project.

Instant Research & Knowledge: Need a quick marketing stat or psychology principle? Ask AI. It gives fast, helpful summaries that save time digging through sources (just double-check facts).

Multitasking for Solo Creators: AI helps you do more — writing, storyboarding, analyzing data — even if you're a one-person team. Great for freelancers or editors wearing many hats.

Cons: What to Watch Out For

Hallucinations (False Info): AI can *sound* smart but make things up. Always fact-check claims like "voted #1" or "used by 80% of dermatologists" — if false, it could damage your brand or be legally risky.

Generic or Off-Brand Tone: Without clear prompts, AI might write cheesy, bland copy that doesn't match your brand voice. Always edit and guide it — it's a first draft, not a final one.

Legal & Ethical Concerns: AI-generated visuals, voices, or text can bring copyright issues. Never use AI to mimic a brand, celebrity, or artist without permission. Don't share confidential info in public AI tools.

Lack of Human Nuance: AI doesn't *feel* — it follows patterns. That means it might miss emotional depth, cultural context, or originality. You still need to bring the human insight that makes ads resonate.

Bias Risks: AI models learn from the internet — which can include stereotypes. Left unchecked, they might generate visuals or ideas that lack diversity or reflect unconscious bias. Always review critically.

Overreliance = Skill Atrophy: Use AI to support your skills, not replace them. Keep practicing hooks, copy, storytelling, and strategy. Don't let AI become a crutch or your work will start sounding like everyone else's.

Tool Limitations & Glitches: Sometimes AI makes mistakes: extra fingers in an image, weird phrasing, or broken logic in a chart summary. Plus, some tools have paywalls or usage caps. You'll still need to verify, edit, and polish.

AI is a superpower — but only if you stay in control. Treat it like a helpful junior teammate: fast, tireless, and creative, but still needs your judgment, editing, and leadership to produce truly great work.

Limitations and Responsible Use of AI

AI is a powerful tool — but it's still just that: a tool. Here's how to use it wisely.

AI is Your Assistant, Not Your Boss

AI gives suggestions, not decisions. It can help you brainstorm or analyze, but you're still the one who decides what fits your brand and audience. Don't blindly follow its advice — *you're the creative director.*

Always Review for Content Safety

AI doesn't understand nuance, current events, or brand context. Double-check everything it creates — text, images, or audio — for tone, accuracy, and unintended issues. If it's going public, put human eyes on it.

Watch What You Input

Don't paste sensitive info like unreleased campaigns or customer data into public AI tools. Use private, secure platforms if needed — and always follow your company's privacy policies.

AI Can't Make a Whole Ad (Yet)

AI is great at scripts, voices, and rough visuals. But it can't shoot real footage, record authentic testimonials, or handle final production. Think of it as a co-pilot for concepting and polishing, not a replacement for your creative process.

Laws Are Changing

AI content (especially voices and images) comes with legal gray areas. Stay informed. Some platforms may require AI disclosure in the future. And remember: cloning someone's voice or likeness without permission is a *hard no.*

Prompting Takes Practice

Getting great results takes a bit of learning. Small tweaks in your prompt can make a huge difference. Be ready to experiment — this is a new creative skill that pays off fast.

Use It to Amplify, Not Replace

Let AI handle the grunt work so you can focus on creativity, storytelling, and strategy — the parts only humans can do. Always polish its outputs and respect creative ownership. AI might get you 80% there — but the last 20% is where your craft shines.

AI Isn't Replacing You — It's Unleashing You

AI is here — not as a threat, but as a powerful sidekick. Think of it as your creative assistant: fast, smart, and available 24/7. You're still the one in control — the strategist, the storyteller, the decision-maker. AI just helps you get there faster.

Yes, the tools can feel overwhelming at first. But you don't need to learn them all overnight. Start small:

- Use **ChatGPT** or **Claude** to rewrite ad copy.

- Try **Midjourney** to generate a moodboard.
- Play with **ElevenLabs** for temp voiceovers.

The more you experiment, the more you'll discover where AI fits into your workflow.

Just like the shift from analog to digital editing, this is the next evolution in creative production. Editors who lean into AI will find they can produce more ideas, iterate faster, and work more strategically. Editors who don't? They may fall behind.

But remember: **AI doesn't have instincts. It doesn't feel.** That's where you come in. Your taste, empathy, humor, and creativity are what connect with real people. AI just helps remove the friction so you can focus on the good stuff — the part only *you* can do.

Embrace AI. Use it to automate the boring parts. Save time. Explore new directions. Build a custom "prompt playbook" just like you'd build a go-to edit flow. Get better, faster, and more confident with each project.

You're not just keeping up — you're leading. You're at the center of a creative shift. AI is here to help you hook, edit, and convert more powerfully than ever.

The future of paid social isn't man *vs.* machine — **it's man *plus* machine**. And you're ready for it.

Citations

1. Influencer Marketing Hub (2024). "Influencer Marketing Benchmark Report."
2. Meta Business Help Center (2023). "Creative Best Practices."
3. AdEspresso by Hootsuite (2023). "How Much Text Should Be on Facebook Ads?"
4. Facebook for Business (2022). "Ad Text Policy and Guidelines."
5. HubSpot (2024). "The Power of UGC in Paid Social Campaigns."
6. Stackla (2022). "Consumer Content Report: Influence in the Digital Age."
7. Buffer (2023). "State of Remote Work and Burnout."
8. Nielsen Norman Group (2022). "F-Shaped Pattern of Reading on the Web."
9. Instapage (2023). "How to Optimize Your CTA for Conversions."
10. 3M Corporation (2019). "Visual Attention in Marketing: Processing Images 60,000x Faster Than Text."
11. Diamond Group (2023). "How Engagement Affects CPC and CPA in Facebook Ads."
12. LinkedIn (2023). "How Audience Insights Drive Creative Decisions."
13. Pathlabs (2022). "Understanding CTR, CVR and ROAS."
14. Linx Digital (2023). "Why CTR Is Still Your Most Important Creative Metric."
15. Wordstream (2023). "CPC and CPA Definitions for Paid Advertising."
16. DashtThis (2023). "Engagement Rate Formulas in Paid Social."
17. Everyonesocial (2023). "Why UGC Still Performs Better Than Branded Ads."

18. Backlinko (2023). "93% of Marketers Say UGC Outperforms Traditional Content."
19. Usetwirl (2022). "Why 85% of Videos Are Watched on Mute (and Why That Matters for Editors)."
20. ClickUp Blog (2023). "Project Management for Creatives: ClickUp vs Asana vs Monday."
21. Asana.com (2024). "Creative Workflow Features."
22. Monday.com (2024). "Creative Use Cases for Agencies."
23. Air.inc (2023). "Organizing Video Review and Delivery for Teams."
24. Meta Ad Library (2024). "Dr. Squatch Sponsored Ad Example." Library ID: 9218317591612937. Retrieved from https://www.facebook.com/ads/library

About the Author

Arie Eskinazi is a paid social video editor, creative strategist, and founder of **VinciVision**, a creative agency specializing in performance-driven advertising. Born and raised in Venezuela, Arie developed an early passion for marketing and storytelling through his family's roots in the advertising world. After immigrating to the United States, he combined hands-on experience with academic training, earning his **Bachelor's Degree in Multimedia Studies** from **Florida Atlantic University**.

Throughout his career, Arie has crafted video ads for a wide range of brands—from local businesses to global giants like **Versace**, **Nike**, **H&M**, **iRestore**, **Noom**, **Fabletics**, **Dr. Squatch**, and many more. His expertise was further sharpened during his time at **TubeScience**, one of the largest paid social agencies in the U.S., where he learned to engineer high-performing ads at scale.

Today, through VinciVision, Arie helps brands grow by blending creative artistry with the science of performance marketing. His approach draws inspiration from **Leonardo da Vinci** himself: balancing creativity with data, imagination with results.

In *Hook. Edit. Convert.*, Arie distills years of agency, brand, and freelance experience into a practical, real-world guide for the next generation of paid social editors, creative producers, and performance-driven creatives.

www.ingramcontent.com/pod-product-compliance
Lightning Source LLC
Chambersburg PA
CBHW041209220326
41597CB00030BA/5142